Springer Proceedings in Mathematics & Statistics

Volume 428

This book series features volumes composed of selected contributions from workshops and conferences in all areas of current research in mathematics and statistics, including data science, operations research and optimization. In addition to an overall evaluation of the interest, scientific quality, and timeliness of each proposal at the hands of the publisher, individual contributions are all refereed to the high quality standards of leading journals in the field. Thus, this series provides the research community with well-edited, authoritative reports on developments in the most exciting areas of mathematical and statistical research today.

Amine Laghrib · Lekbir Afraites ·
Mourad Nachaoui
Editors

New Trends of Mathematical Inverse Problems and Applications

ICNTAM 2022, Béni Mellal, Morocco,
May 19–21

Springer

Editors
Amine Laghrib
Department of Mathematics
Université Sultan Moulay Slimane
Béni Mellal, Morocco

Lekbir Afraites
Department of Mathematics
Université Sultan Moulay Slimane
Béni Mellal, Morocco

Mourad Nachaoui
Department of Mathematics
Université Sultan Moulay Slimane
Béni Mellal, Morocco

ISSN 2194-1009 ISSN 2194-1017 (electronic)
Springer Proceedings in Mathematics & Statistics
ISBN 978-3-031-33068-1 ISBN 978-3-031-33069-8 (eBook)
https://doi.org/10.1007/978-3-031-33069-8

Mathematics Subject Classification: 65K10, 90C26, 68U10

This Springer imprint is published by the registered company Springer Nature Switzerland AG
The registered company address is: Gewerbestrasse 11, 6330 Cham, Switzerland

Preface

This book collects selected papers presented at the "International Conference on New Trends of applied Mathematics ICNTAM", which is held at Béni Mellal, Morocco, from 19 to 21 May 2022. The book is focused on topics related to "New Trends of Mathematical Inverse Problems and Applications". The evolution of mathematics gave rise to several new fields, among them we can find Data Science and some applications to mathematics and computer science. Recently, Data Science is considered as a dynamic and attractive research area with numerous usability in other scientific fields, such as: Machine Learning, Data Mining, and computer science, especially in Image Processing. For that and to closely follow the evolution of Data Science using recent mathematical tools, we propose this book which contains a number of papers on various mathematical tools, including inverse problems, optimization, and their applications.

All contributing authors are eminent researchers and scholars in their respective fields, hailing from around the world. The book has a great income for researchers, scientists, and engineers from both academia and industry.

Béni Mellal, Morocco

Amine Laghrib
Lekbir Afraites
Mourad Nachaoui

Contents

Comparing Numerical Methods for Inverse Source Problem in Time-Fractional Diffusion Equation

A. Oulmelk, M. Srati, and L. Afraites

Abstract The purpose of this work is to estimate a source term in the time-fractional diffusion equation from additional measurements. After recasting the inverse source problem as an nonsmooth optimization problem. Two different alternative methods are applied to find the solution of the control problem. The accuracy of the proposed methods is tested in the cases of with and without noise, and the findings are very promising and encouraging.

Keywords Inverse source problem · Time-fractional diffusion equation · Optimal control problem · Descent gradient method and Primal-Dual method

1 Introduction

The fractional diffusion equation is a generalization of the classical diffusion equation by replacing the standard time derivative by a time-fractional which models anomalous diffusive phenomena [1, 2]. In recent years, inverse problems for time-fractional diffusion equation have become very active in interdisciplinary research area. The fractional diffusion equation has been introduced in physics to describe diffusions in media with fractal geometry. The authors in [5] pointed out that field data show anomalous diffusion in a highly heterogeneous aquifer. Also in [6] they applied the continuous-time random walk for better simulations for the anomalous diffusion in an underground environmental problem. One can regard (1) as a macroscopic model derived from the continuous-time random walk, and in [7] they demonstrated that a fractional diffusion equation describes a non-Markovian diffusion process with a memory. In [8] they pointed out that the fractional wave equation governs the propagation of mechanical diffusive waves in viscoelastic media.

A. Oulmelk (✉) · M. Srati · L. Afraites
EMI FST Beni-Mellal, Sultan Moulay Slimane University, Beni Mellal, Morocco
e-mail: oulmelkabdessamad@gmail.com

L. Afraites
e-mail: l.afraites@usms.ma

A. Laghrib et al. (eds.), *New Trends of Mathematical Inverse Problems and Applications*, Springer Proceedings in Mathematics & Statistics 428,
https://doi.org/10.1007/978-3-031-33069-8_1

In this paper we study an inverse space-dependent source problem for a time-fractional diffusion equation in a bounded domain. Let Ω be a bounded domain in $\mathbb{R}$ with sufficiently smooth boundary $\partial\Omega$ and $I = (0, T)$ where $T > 0$ is the final time. Considering the following time-fractional diffusion equation

$$\begin{cases} D_t^\alpha u(x,t) - \Delta u(x,t) = G(x,t) + f(x), & (x,t) \in \Omega \times I, \\ u(x,0) = \phi(x), & x \in \Omega, \\ u(x,t) = 0, & (x,t) \in \partial\Omega \times I, \end{cases} \tag{1}$$

where ϕ is the initial data.

This paper is dedicated to the estimation of f in the problem (1) using the following final time measures

$$u(x,T) = h(x), \quad a.e. \ \ x \in \Omega, \tag{2}$$

where h is a given function.

In the Eq. (1), the notation D_t^α is called the Caputo fractional derivative of order $\alpha \in (0, 1)$ with respect to time and it's defined by [9, 10]

$$D_t^\alpha u(t) = \frac{1}{\Gamma(1-\alpha)} \int_0^t \frac{u'(s)}{(t-s)^\alpha} ds, \quad t > 0,$$

where $\Gamma(.)$ denotes Euler's Gamma function. In the case $\alpha = 1$, $D_t^\alpha u$ is identified with the usual first order derivative $\partial_t u$.

There are many works on the mathematical treatments of Eq. (1) concerning the existence, uniqueness and regularity of the solution. For example, in [11], the principle of maximum solution is proved. While, the [12] presented some results for existence and uniqueness. Similarly, the work [3, 4, 13, 14] showed the existence, uniqueness and regularity of the solution by means of Mittag-Leffler functions.

The inverse source problem mentioned above has been examined in a number of studies. For example, the authors in [15] used the regularized control optimal approach under two different cost functions to analyse an inverse source problem for a time-fractional diffusion equation, and in [16, 17] they solved an inverse source problem for a time-fractional diffusion equation on the basis of the optimal control method. Also in [18] they investigated an inverse time-dependent source problem for a multi-dimensional fractional diffusion equation from boundary Cauchy data. While in [19] considered an inverse space-dependent source problem by using the Cauchy data at one end $x = 0$, and in [20] they used a reproducing kernel space method to solve an inverse space-dependent source problem from the final data. As in [21] they proposed a modified quasi-boundary value method for identifying the space-dependent source term by using the final data, and in [22, 23] they solved an inverse space-dependent source problem for a space-time fractional diffusion equation. For the integer-order cases, the inverse source problems have been widely studied, see [24–26]. In the case $\alpha = 1$, the inverse source problem for the model (1) has also been analyzed in several works [27–30].

We consider an optimal control approach for the inverse problem (1)–(2) based on the L^1 fitting, which adapts when the data h is not regular as the most practical data in general, and formulated as follows:

$$\min_{f \in \mathcal{U}_{ad}} \left\{ J(f) = \frac{1}{2} \int_\Omega |u(x,T) - h(x)| dx + \frac{\beta}{2} \int_\Omega |f|^2 dx \right\}. \tag{3}$$

where $u(x, T)$ is the solution of the direct problem (1) at final time T, $\beta > 0$ is a regularization parameter (see [31, 32, 40–42, 45, 46] for the choice of β) and $\mathcal{U}_{ad}$ is the admissible set defined by

$$\mathcal{U}_{ad} = \left\{ f \in L^2(\Omega); |f| \leq a \quad a.e. \, \Omega \right\},$$

where a is a positive number. In the literature, various methods are studied to solve the nonsmooth optimization problem [33]. The well-posedness of the nonsmooth optimal control problem (3) has been proven in our previous work [15].

The main issue of this paper is to estimate the source term in a time-fractional diffusion equation using observational data. To this end, we propose two different alternative numerical methods that are based on the nonsmooth optimal control problem (3) constrained by the time-fractional diffusion equation (1). The first proposed method is Primal-Dual, and the second is Descent-Gradient. We also compare and contrast these two different algorithms.

This work's structure is as follows: We use the Primal-Dual method to solve control problem (3) in the next section. The Descent-Gradient algorithm is outlined in Sect. 3. In the one-dimensional case, the numerical results for some examples are presented in Sect. 4. In Sect. 5, we present the conclusion of the paper.

2 Primal-Dual Method

In this section, we propose an effective Primal-Dual method [33, 34] that is well suited to the fractional problem, which introduces two operators. In fact, we convert the nonsmooth optimal control problem under study into the following kind of optimization problem:

$$\widehat{f} = \arg\min_f \{ \mathrm{L}(K(f)) + \mathrm{N}(f) \}, \tag{4}$$

where K is a nonlinear operator that involves solving the PDE-fractional equation (1), L and N are proper, convex, and lower semicontinuous functionals and L is a nonsmooth discrepancy.

Firstly, The operator $\mathcal{K}$ that maps f to the solution u of the PDE-fractional in (1) at final time T is constructed

$$\begin{aligned} \mathcal{K}\ : \mathcal{U}_{ad} &\longrightarrow L^2(\Omega) \\ f &\longmapsto u(T,x), \end{aligned}$$

The operators L, N, K_1 and K_2 are now defined as follows. $\mathrm{L}(z_1, z_2) = \|z_1\|_{L^1} + \frac{\beta}{2}\|z_2\|_{L^2}^2$, $K_1(f) = \mathcal{K}(f) - h$, $K_2(f) = f$, and $K(f) = \begin{pmatrix} K_1(f) \\ K_2(f) \end{pmatrix}$ such as

$$\mathrm{L}(K(f)) = \left\|\mathcal{K}(f) - h\right\|_{L^1} + \frac{\beta}{2}\left\|f\right\|_{L^2}^2,$$

and

$$\mathrm{N}(f) = i_{\mathcal{U}_{ad}}(f),$$

is the indicator function of a subset $\mathcal{U}_{ad}$ of a set $L^2(\Omega)$.
The above notations transform the optimization problem (3) into

$$\inf_f \left\{\mathrm{L}(K(f)) + \mathrm{N}(f)\right\}, \tag{5}$$

This is how the general saddle-point problem is presented.

$$\min_f \max_l \mathrm{N}(f) + < K(f), l > -\mathrm{L}^*(l), \tag{6}$$

where L^* is the Fenchel conjugate function of f, which is defined as follows

$$\mathrm{L}^*(f) = \sup_x < x, f > -\mathrm{L}(x).$$

We may use the calculus of Clarke's generalized derivative (See [35]) to determine for (4) the overall system of critical point conditions for convex operators L and N and continuously differentiable K.

$$\begin{cases} K(\hat{f}) \in \partial \mathrm{L}^*(\hat{l}), \\ -[K'(\hat{f})]^*\hat{l} \in \partial \mathrm{N}(\hat{f}), \end{cases} \tag{7}$$

where $[K'(\hat{f})]^* = \Big([K_1'(\hat{f})]^*, [K_2'(\hat{f})]^*\Big)$, $\hat{l} = \begin{pmatrix} \hat{l}_1 \\ \hat{l}_2 \end{pmatrix}$ and $[K'(\hat{f})]^*\hat{l} = [K_1'(\hat{f})]^*\hat{l}_1 + [K_2'(\hat{f})]^*\hat{l}_2$. Moreover, for a given $\hat{l}_1$, $[K_1'(\hat{f})]^*\hat{l}_1$ is the directional adjoint derivative at $\hat{f}$, it was calculated in [15] by

$$\left(K_1'(\hat{f})\right)^*\hat{l}_1 = \int_0^T v(x,t)dt, \tag{8}$$

where v is the solution to the adjoint problem below

$$\begin{cases} D_T^{\alpha} v(x,t) - \Delta v(x,t) = \hat{l}(x), & (x,t) \in \Omega \times I, \\ v(x,T) = 0, & x \in \Omega, \\ v(x,t) = 0, & (x,t) \in \partial\Omega \times I. \end{cases} \tag{9}$$

We are now developing the first-order optimality conditions (7). When we apply the Primal-Dual algorithm iterations to these conditions [36], we get the iterative method shown below.

$$\begin{cases} f^{n+1} = \mathsf{prox}_{\tau \mathrm{N}}\big(f^n - \tau (K'(f^n))^* l^n\big), \\ \widehat{f}^{n+1} = f^{n+1} + \theta(f^{n+1} - f^n), \\ l^{n+1} = \mathsf{prox}_{\sigma \mathrm{L}^*}\big(l^n + \sigma K(\widehat{f}^{n+1})\big), \end{cases} \tag{10}$$

then we apply the accelerated version of Primal-Dual [37] given in the algorithm 1. We must define the proximal operators $\mathsf{prox}_{\tau \mathrm{N}}$ and $\mathsf{prox}_{\sigma \mathrm{L}^*}$ in order to take use of the iteration method. The first operator is defined as follows.

$$f = \mathsf{prox}_{\tau \mathrm{N}}(\bar{f}) = arg \min_z \frac{\|z - \bar{f}\|^2}{2\tau} + N(z),$$

since $N(.) = i_{\mathcal{U}_{ad}}(.)$ is differentiable, direct calculation yields the following

$$f(x) = \mathsf{prox}_{\tau \mathrm{N}}(\bar{f})(x) = P_{\mathcal{U}_{ad}}(\bar{f}) = \max(\min(-a, \bar{f}), a). \tag{11}$$

We now elaborate the second operator. $\mathsf{prox}_{\sigma \mathrm{L}^*}(.)$. Suppose that $\mathrm{L}(z_1, z_2) = \mathrm{L}_1(z_1) + \mathrm{L}_2(z_2)$, where L_1 and L_2 are two operators that are defined differently by

$$\begin{aligned} \mathrm{L}_1 \ : L^1(\Omega) &\longrightarrow \mathbb{R}^+ \\ z_1 &\longmapsto \|z_1\|_{L^1}, \end{aligned}$$

$$\begin{aligned} \mathrm{L}_2 \ : L^2(\Omega) &\longrightarrow L^2(\Omega) \\ z_2 &\longmapsto \frac{\beta}{2}\|z_2\|^2_{L^2}. \end{aligned}$$

Next, we have $\mathrm{L}^* = \mathrm{L}_1^* + \mathrm{L}_2^*$ and $\mathsf{prox}_{\sigma \mathrm{L}^*}\begin{pmatrix} \hat{l}_1 \\ \hat{l}_2 \end{pmatrix} = \begin{pmatrix} \mathsf{prox}_{\sigma \mathrm{L}_1^*}(\hat{l}_1) \\ \mathsf{prox}_{\sigma \mathrm{L}_2^*}(\hat{l}_1) \end{pmatrix}$.

Utilizing Moreau's identity makes computing the proximal operator of the dual function L_1^*:

$$l_1 = \mathsf{prox}_{\sigma \mathrm{L}_1^*}(\hat{l}_1) = \hat{l}_1 - \sigma \mathsf{prox}_{\sigma^{-1}\mathrm{L}_1}(\sigma^{-1}\hat{l}_1). \tag{12}$$

According to the definition of the Fenchel conjugate function, direct calculation yields the following

$$l_2 = \mathsf{prox}_{\sigma \mathrm{L}_2^*}(\hat{l}_2) = \left(\frac{\beta}{\beta + \sigma}\right)\hat{l}_2. \tag{13}$$

Currently, Algorithm 1 describes the main algorithm of our approach to build the Primal-Dual algorithm connected to the problem (4).

Algorithm 1 Accelerated Primal-Dual algorithm

Given a second member F, initial function ϕ, noisy data h and the precision tol .
Choose $\mu > 0$, α, $T > 0$, $\tau_0 > 0$, $\sigma_0 > 0$, initialization of variable f^0 and vector l^0
The reconstruction of the parameter component f

1. $f^{n+1} = \text{prox}_{\tau_n \text{N}}(f^n - \tau_n[K'(f^n)]^* l^n)$ Use (11) for $\text{prox}_{\tau_n \text{N}}$
2. $\theta_n = \frac{1}{\sqrt{1+2\mu\sigma_n}}$, $\tau_{n+1} = \frac{\tau_n}{\theta_n}$ and $\sigma_{n+1} = \theta_n \sigma_n$.
3. $z^{n+1} = f^{n+1} + \theta_n(f^{n+1} - f^n)$
4. $l^{n+1} = \text{prox}_{\sigma_n \text{L}^*}(l^n + \sigma_{n+1} K(z^{n+1}))$ Use (12) and (13) for $\text{prox}_{\sigma \text{L}^*}$

Iterate Until $Err(f_{m+1}) < tol$. Use (19) for Err

---

3 Descent-Gradient Method

A classical way to solve an optimization problem is to use Descent-Gradient methods. However, the fidelity term in the optimal control problem (3) is not differentiable, hence one may utilize optimization methods, such as the sub-gradient method, which does not need any differentiability assumption on the cost functional. However, these methods are very expensive in terms of computation time [38, 39]. Therefore, we opt for an approximation of the fidelity term by a regular function that is twice differentiable and convex. This allows us to use Descent-Gradient methods to solve the resulting optimal control problem.

The cost functional $J(f)$ can be approximated by a functional $J_\varepsilon(f)$ which is differentiable and convex, as follows:

$$J_\varepsilon(f) = \frac{1}{2} \int_\Omega \left(|u(x,T) - h(x)|^2 + \varepsilon^2\right)^{\frac{1}{2}} dx + \frac{\beta}{2} \int_\Omega |f|^2 dx, \tag{14}$$

where $\varepsilon > 0$ is a positive parameter. We have that J_ε converges to J, when ε tends to zero. This is illustrated in in the Fig. 1.

The approximate optimal control problem then became as follows

$$\bar{f} = \arg \min_{f \in \mathcal{U}_{ad}} J_\varepsilon(f), \tag{15}$$

The existence and uniqueness results of the solution for above optimal control problem are investigated in [16]. In order to illustrate the solution to the approximate optimal control problem (15), we calculate the gradient related to the cost function using the adjoint state system.

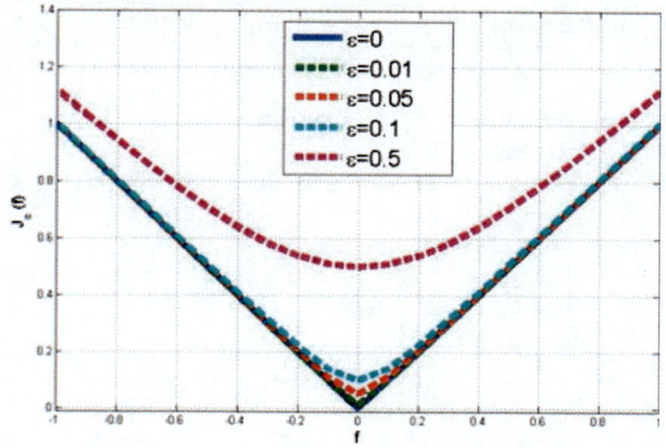

Fig. 1 Regularized cost functional $J_\varepsilon(f)$ for different parameters ε

Doing this requires, we note that for any $f \in \mathcal{U}_{ad}$ and perturbation $df \in L^2(\Omega)$, $u^1 = u^1(f, df)$ the sensitivity of the state satisfying the following system

$$\begin{cases} D_t^\alpha u^1(x,t) - \Delta u^1(x,t) = df(x), & (x,t) \in \Omega \times I = Q, \\ u^1(x,0) = 0, & x \in \Omega, \\ u^1(x,t) = 0, & (x,t) \in \partial\Omega \times I. \end{cases} \tag{16}$$

The sensitivity problem (16) allows for the calculation of the gradient of the cost functional (14). This finding was presented in our prior article [15]. The following proposition demonstrates the functional cost gradient

Proposition 1 *The gradient of the cost function J_ε is provided by*

$$J'_\varepsilon(f) = \frac{1}{2}\int_0^T v(x,t)dt + \beta f(x), \tag{17}$$

where v is the solution of the adjoint problem

$$\begin{cases} D_T^\alpha v(x,t) - \Delta v(x,t) = \left((u(x,t) - h(x))^2 + \varepsilon^2\right)^{-\frac{1}{2}} (u-h)\delta(t-T), & (x,t) \in \Omega \times I, \\ v(x,T) = 0, & x \in \Omega, \\ v(x,t) = 0, & (x,t) \in \partial\Omega \times I, \end{cases} \tag{18}$$

and δ is the Dirac function.

After calculating the gradient of the cost functional J_ε, it is now possible to provide the Descent-Gradient algorithm for he approximate optimal control problem (15).

Algorithm 2 Descent-Gradient.

Input: the error tol, the initial condition u_0, the data h, the second member G, α , β and ς^0.

1. **Initialize**: $f^0 \in \mathcal{U}_{ad}$ and set $n = 0$.
2. **Repeat**
3. Compute u^n by solving the direct problem (1).
4. Compute v^n by solving the adjoint problem (18).
5. Compute $J_\varepsilon'(f^n)$ by (17).
6. **Update** $f^{n+1} = f^n - \varsigma^n J_\varepsilon'(f^n)$ with ς^n is computed by Armijo line search.
7. **Until** $\| J\varepsilon'(f^{k+1})\| \leq tol$.

4 Numerical Results

In this section, we provide several examples for the one-dimensional case to evaluate the efficiency of the suggested methods and compare them. By setting the unknown source term f and selecting the second member function $G(x, t)$ as the initial condition $\phi(x)$, we construct the synthetic data $h(x)$. We then solve the direct problem (1) using the finite difference method developed in [43, 44], we extract the measurement $h(x)$ as the solution $u(x, T)$. In numerical computations, we always set $T = 1$ and $\Omega = (0, 1)$. The noisy is generated by adding a random perturbation, i.e.

$$h^\mu = h + \mu h.(2.rand(size(h) - 1).$$

$\mu = \|h^\mu - h\|_{L^2(\Omega)}$ determines the equivalent noise level. We compute the error represented below to show the accuracy of the numerical solution of the suggested methods.

$$e_m = \| f_m(x) - f(x)\|_{L^2(\Omega)},$$

where $f(x)$ is the exact solution and $f_m(x)$ is the source term that was reconstructed at the m-th iteration.

The application also needs a stopping condition. In our numerical studies, it was challenging to identify criteria that are applicable to test data that is both synthetic and measured. Since the change in the reconstruction obtained from each iteration step in our case was no longer discernible, the iterative process was interactively monitored and stopped. Then is utilized the calculated relative error presented by

$$Err(f_{m+1}) = \frac{e_m}{\|f(x)\|_2}. \tag{19}$$

In this numerical experiment, we assume that the solution of the direct problem to be $u(x, t) = t^2 \sin(\pi x) + \sin(\pi x)$, the order of the fractional derivative to be $\alpha = 0.5$ and the regularity parameter $\beta = 10^{-6}$. We now provide four examples for the one-dimensional case to demonstrate the effectiveness of the suggested methods.

Example 1 Suppose $f(x) = \pi^2 \sin(\pi x)$ will be identified and the final data $u(x, T)$ are obtained by solving the direct problem (1) with $G(x, t) = \left(\frac{2t^{2-\alpha}}{\Gamma(3-\alpha)} + (\pi t)^2\right)$ $\sin(\pi x)$ and initial data $\phi(x) = \sin(\pi x)$.

Example 2 In the second example, which is synthetic, we use the values of G and ϕ from the previous example as well as the following assumption about the unknown source term:

$$f(x) = (1 + x^2). \tag{20}$$

By using the finite difference method to solve the direct problem (1), the final data $u(x, T)$ are obtained.

Example 3 In this example, a non-smooth example with a peak is tested. Assume that the source function in the following expression is given.

$$f(x) = \begin{cases} x & 0 \le x \le \frac{1}{2} \\ 1 - x & \frac{1}{2} \le x \le 1 \end{cases} \tag{21}$$

with $G(x, t) = \left(\frac{2t^{2-\alpha}}{\Gamma(3-\alpha)} + (\pi t)^2\right) \sin(\pi x)$ and initial data $\phi(x) = \sin(\pi x)$. By using the exact input data as the finite difference method to solve the direct problem (1), the final data $u(x, T)$ is obtained.

Example 4 As a last example, let's have a look at a discontinuous source function with the data of the first example. Suppose

$$f(x) = \begin{cases} 0 & 0 \le x \le \frac{1}{3} \\ 1 & \frac{1}{3} < x \le \frac{2}{3} \\ 0 & \frac{2}{3} < x \le 1. \end{cases} \tag{22}$$

4.1 *Primal-Dual Algorithm*

This section looks into the results of the first algorithm, concerned Primal-Dual method. For Examples 1 and 2, we compare the exact source in Fig. 2 with the reconstructed source. Examples 3 and 4 identify source terms shown in Fig. 3. The advantage of this method is shown by the fact that the constructions work well in both regular and singular cases. By presenting the identified source along with the noise in Fig. 4, we demonstrate the stability of the Primal-Dual method.

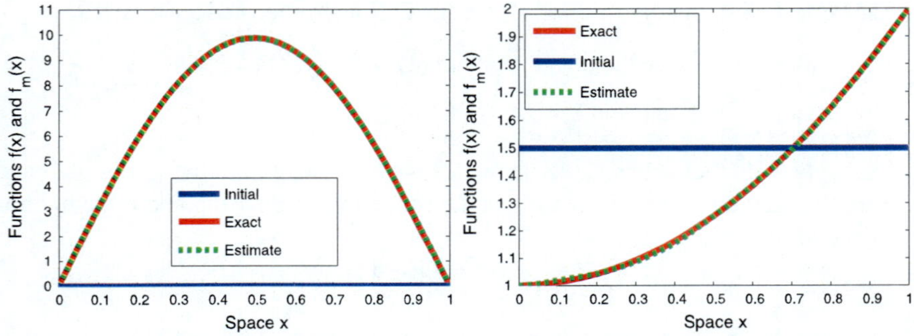

(a) Example 1 without noise (b) Example 2 without noise

Fig. 2 The numerical output of the Primal-Dual algorithm for the Examples 1 and 2

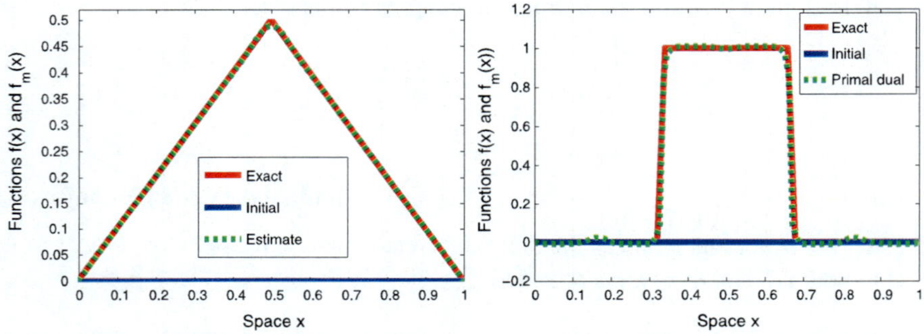

(a) Example 3 without noise (b) Example 4 without noise

Fig. 3 The numerical output of the Primal-Dual algorithm for the Examples 3 and 4

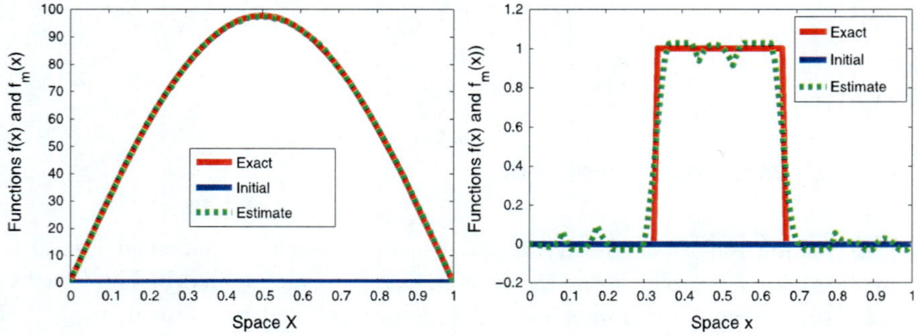

(a) Example 1 with 5% of noise (b) Example 4 with 5% of noise

Fig. 4 The numerical results of the Primal-Dual algorithm with noise

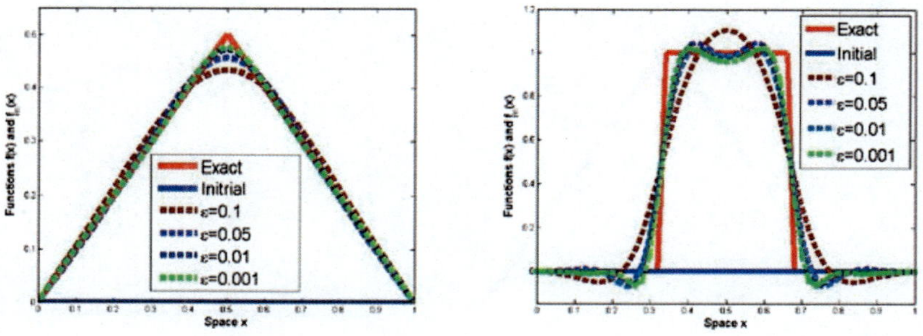

(a) Example 3 (b) Example 4

Fig. 5 The numerical output of the Descent-Gradient algorithm for different ε

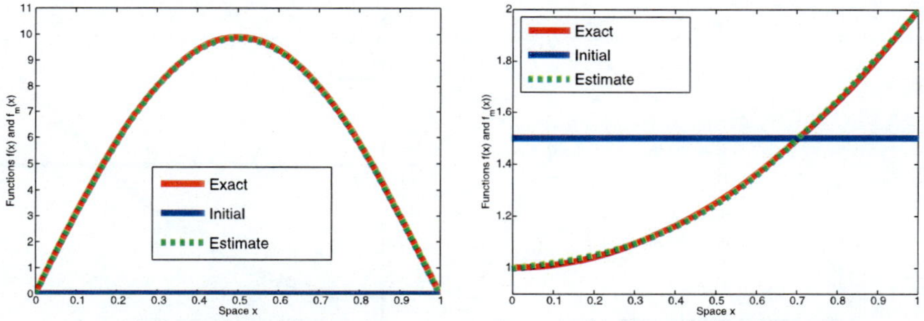

(a) Example 1 without noise (b) Example 2 without noise

Fig. 6 The numerical output of the Descent-Gradient algorithm for noise-free regular examples

4.2 Descent-Gradient Algorithm

In this subsection, we show the numerical output corresponding to the Descent-Gradient Algorithm 2. We also show the effectiveness of the Descent-Gradient method by presenting the numerical results of estimating the source function f for the past four examples.

Firstly, we provide the numerical outcomes of the source term reconstruction for different ε. Figure 5 shows that the numerical results obtained from the Descent-Gradient method are close to coinciding with those obtained from the Primal-Dual method when ε tends to zero.

We begin with the regular examples, such as Examples 1 and 2, the results of which are shown in Fig. 6. We note that the results obtained in these typical cases virtually match the exact source. On the other hand, with complex sources like Examples 3 and 4, the method does not precisely construct the singularities of the source Fig. 7.

When we add noise, we present the numerical results that were drawn from the noisy data. For regular cases, Fig. 8 shows the outcomes with noise, whereas Fig. 9 shows the outcomes for complex cases with noise.

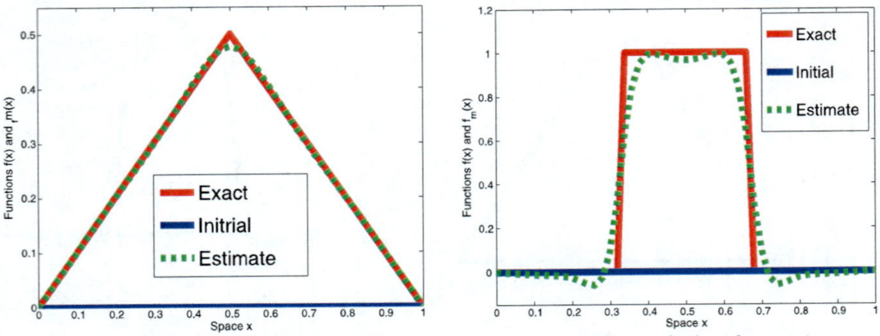

(a) Example 3 without noise (b) Example 4 without noise

Fig. 7 The numerical output of the Descent-Gradient algorithm without noise for Examples 3 and 4

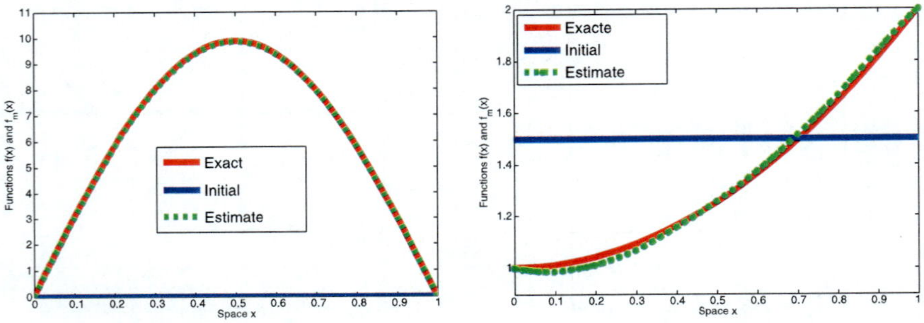

(a) Example 1 with 5% of noise (b) Example 2 with 5% of noise

Fig. 8 The numerical output of the Descent-Gradient algorithm for regular examples with noise

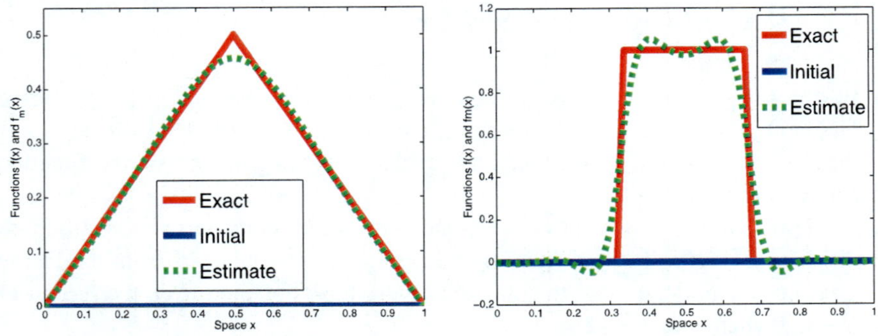

(a) Example 3 with 5% of noise (b) Example 4 with 5% of noise

Fig. 9 The numerical output of the Descent-Gradient algorithm with noise

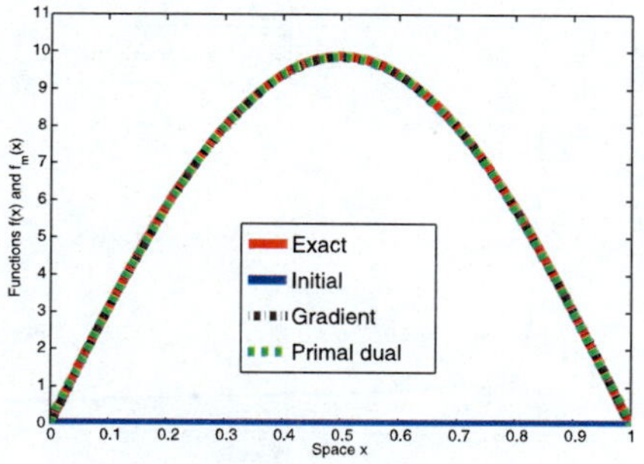

(a) Example 1 without noise

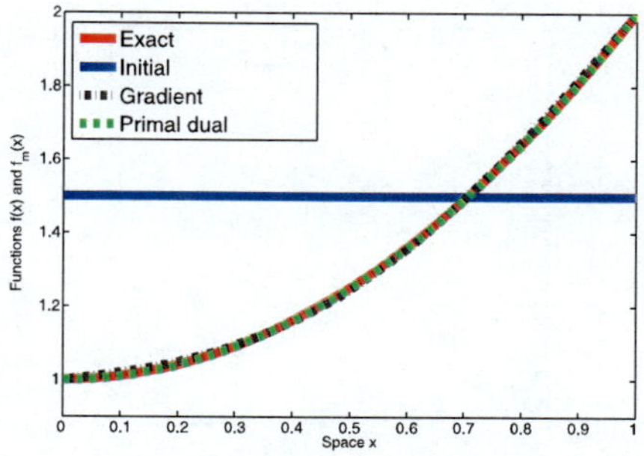

(b) Example 2 with 5% of noise

Fig. 10 The numerical results of comparison without noise

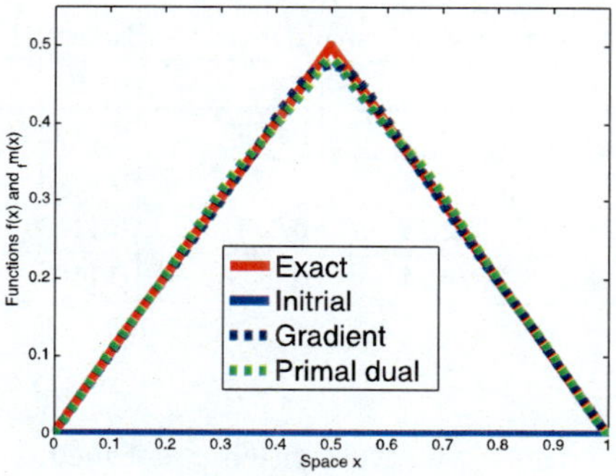

(a) Example 3 without noise

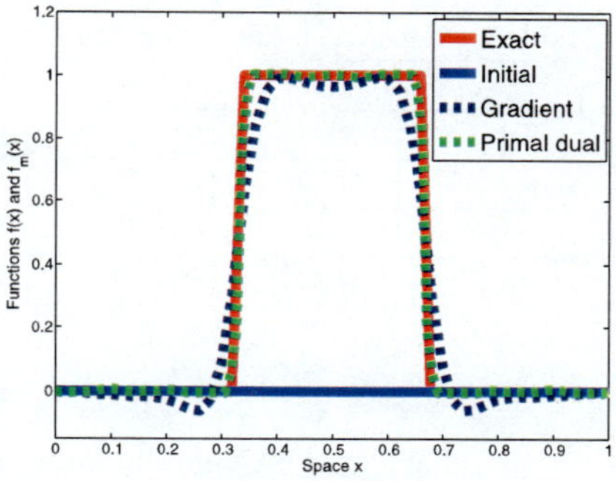

(b) Example 4 without noise

Fig. 11 The numerical results of comparison with noise

The Descent-Gradient method is stable with respect to the noise, based on the numerical experiments with noise, but it is unable to capture the singularity of the sources.

4.3 Comparison

The next section is devoted to the comparison between the Primal-Dual method and the Descent-Gradient method for all the above examples. Figures 10 and 11 show that while both methods act identically for regular sources. In contrast, the Primal-Dual method outperforms the Descent-Gradient method for the discontinuous examples.

Figure 12 shows the relative error for Example 1 of the two proposed algorithms and shows that the Primal-Dual method allows for a faster convergence rate than the Descent-Gradient method. We can observe that the Primal-Dual method is more accelerated than the Descent-Gradient method.

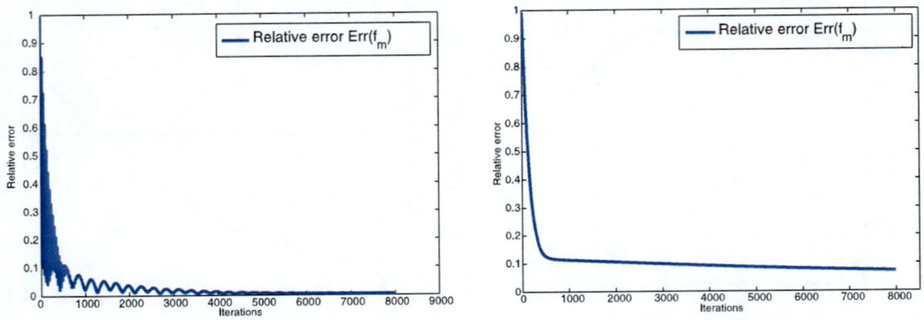

(a) Primal-Dual (b) Descent-Gradient

Fig. 12 Relative errors of two different methods for Example 1

Table 1 Numerical results by Descent-Gradient method for e_m with different μ in Example 1

μ	0.001	0.005	0.01	0.05	0.1	0.15
$\alpha = 0.3$	0.0087	0.0136	0.0293	0.0571	0.0743	0.1102
$\alpha = 0.5$	0.0087	0.0137	0.0295	0.0572	0.0744	0.1104
$\alpha = 0.7$	0.0088	0.0138	0.0295	0.0573	0.0745	0.1103
$\alpha = 0.9$	0.0089	0.0138	0.0296	0.0574	0.0746	0.1107

In Table 1, we provide the numerical errors e_m by Descent-Gradient method of Example 1 for various noise levels μ in $\alpha = 0.3$, $\alpha = 0.5$, $\alpha = 0.7$, $\alpha = 0.9$. We can observe that when the noise levels increase, the numerical errors e_m also increase.

5 Conclusion

In this paper, we study an inverse source problem for a time-fractional diffusion equation on the basis of the nonsmooth optimal control method. In studying it, we proposed two different alternative methods and compared them. In the end, numerical tests are shown to demonstrate the effectiveness of the suggested algorithms. Four numerical examples are given to demonstrate how the Prima-Dual method is more effective than the Descent-Gradient method.

References

1. B. Berkowitz, H. Scher, S.E. Silliman, Anomalous transport in laboratory-scale, heterogeneous porous media. Water Resour. Res. **36**(1), 149–158 (2000)
2. R. Metzler, J. Klafter, The random walk's guide to anomalous diffusion: a fractional dynamics approach. Phys. Rep. **339**(1), 1–77 (2000)

3. A. Oulmelk, M. Srati, L. Afraites, A. Hadri, Implementation of the ADMM approach to constrained optimal control problem with a nonlinear time-fractional diffusion equation. Discrete Contin. Dyn. Syst.-S (2022)
4. A. Oulmelk, L. Afraites, A. Hadri, An inverse problem of identifying the coefficient in a nonlinear time-fractional diffusion equation. Comput. Appl. Math. **42**(1), 65 (2023)
5. E.E. Adams, L.W. Gelhar, Field study of dispersion in a heterogeneous aquifer: 2. spatial moments analysis. Water Resour. Res. **28**(12), 3293–3307 (1992)
6. Y. Hatano, N. Hatano, Dispersive transport of ions in column experiments: an explanation of long-tailed profiles. Water Resour. Res. **34**(5), 1027–1033 (1998)
7. R. Metzler, J. Klafter, Boundary value problems for fractional diffusion equations. Physica A: Stat. Mech. Appl. **278**(1–2), 107–125 (2000)
8. F. Mainardi, Fractional diffusive waves in viscoelastic solids. Nonlinear Waves Solids **137**, 93–97 (1995)
9. B. Jin, *Fractional Differential Equations* (Springer, 2021)
10. A.A. Kilbas, H.M. Srivastava, J.J. Trujillo, *Theory and Applications of Fractional Differential Equations*, vol. 204 (Elsevier, 2006)
11. Y. Luchko, Maximum principle for the generalized time-fractional diffusion equation. J. Math. Anal. Appl. **351**(1), 218–223 (2009)
12. K. Sakamoto, M. Yamamoto, Initial value/boundary value problems for fractional diffusion-wave equations and applications to some inverse problems. J. Math. Anal. Appl. **382**(1), 426–447 (2011)
13. B. Jin, B. Li, Z. Zhou, Numerical analysis of nonlinear subdiffusion equations. SIAM J. Numer. Anal. **56**(1), 1–23 (2018)
14. B. Kaltenbacher, W. Rundell, On the identification of a nonlinear term in a reaction-diffusion equation. Inverse Prob. **35**(11), 115007 (2019)
15. A. Oulmelk, L. Afraites, A. Hadri, M. Nachaoui, An optimal control approach for determining the source term in fractional diffusion equation by different cost functionals. Appl. Numer. Math. **181**, 647–664 (2022)
16. Y.-K. Ma, P. Prakash, A. Deiveegan, Optimization method for determining the source term in fractional diffusion equation. Math. Comput. Simul. **155**, 168–176 (2019)
17. X.B. Yan, T. Wei, Inverse space-dependent source problem for a time-fractional diffusion equation by an adjoint problem approach. J. Inverse Ill-posed Probl. **27**(1), 1–16 (2019)
18. Y.S. Li, T. Wei, An inverse time-dependent source problem for a time-space fractional diffusion equation. Appl. Math. Comput. **336**, 257–271 (2018)
19. Y. Zhang, X. Xu, Inverse source problem for a fractional diffusion equation. Inverse Probl. **27**(3), 035010 (2011)
20. W. Wang, M. Yamamoto, B. Han, Numerical method in reproducing kernel space for an inverse source problem for the fractional diffusion equation. Inverse Probl. **29**(9), 095009 (2013)
21. T. Wei, J. Wang, A modified quasi-boundary value method for an inverse source problem of the time-fractional diffusion equation. Appl. Numer. Math. **78**, 95–111 (2014)
22. S. Tatar, R. Tinaztepe, S. Ulusoy, Determination of an unknown source term in a space-time fractional diffusion equation. J. Fract. Calc. Appl. **6**(1), 83–90 (2015)
23. S. Tatar, S. Ulusoy, An inverse source problem for a one-dimensional space-time fractional diffusion equation. Appl. Anal. **94**(11), 2233–2244 (2015)
24. G. Bao, T.A. Ehlers, P. Li, Radiogenic source identification for the helium production-diffusion equation. Commun. Comput. Phys. **14**(1), 1–20 (2013)
25. K. Sakamoto, M. Yamamoto, Inverse heat source problem from time distributing overdetermination. Appl. Anal. **88**(5), 735–748 (2009)
26. K. Sakamoto, M. Yamamoto, Initial value/boundary value problems for fractional diffusion-wave equations and applications to some inverse problems. J. Math. Anal. Appl. **382**(1), 426–447 (2011)
27. L. Afraites, A. Hadri, A. Laghrib, M. Nachaoui, A non-convex denoising model for impulse and gaussian noise mixture removing using bi-level parameter identification. Inverse Probl. Imaging (2022)

28. L. Afraites, A. Hadri, A. Laghrib, M. Nachaoui, A weighted parameter identification PDE-constrained optimization for inverse image denoising problem. Vis. Comput. **38**(8), 2883–2898 (2022)
29. A. Laghrib, L. Afraites, A. Hadri, M. Nachaoui, A non-convex PDE-constrained denoising model for impulse and gaussian noise mixture reduction. Inverse Probl. Imaging (2022)
30. M. Nachaoui, L. Afraites, A. Hadri, A. Laghrib, A non-convex non-smooth bi-level parameter learning for impulse and gaussian noise mixture removing. Commun. Pure Appl. Anal. **21**(4), 1249 (2022)
31. L. Afraites, A. Atlas, Parameters identification in the mathematical model of immune competition cells. J. Inverse Ill-posed Probl. **23**(4), 323–337 (2015)
32. S. Lyaqini, M. Nachaoui, Identification of genuine from fake banknotes using an enhanced machine learning approach, in *International Conference on Numerical Analysis and Optimization Days* (Springer, 2021), pp. 59–70
33. A. Hadri, M. Nachaoui, A. Laghrib, A. Chakib, L. Afraites, A primal-dual approach for the robin inverse problem in a nonlinear elliptic equation: the case of the $L^1 - L^2$ cost functional. J. Inverse Ill-posed Probl. (2022)
34. S. Lyaqini, M. Nachaoui, A. Hadri, An efficient primal-dual method for solving non-smooth machine learning problem. Chaos Solitons Fractals **155**, 111754 (2022)
35. F.H. Clarke, *Optimization and Nonsmooth Analysis*. Classics in Applied Mathematics (SIAM, Philadelphia, 1990)
36. A. Chambolle, T. Pock, A first-order primal-dual algorithm for convex problems with applications to imaging. J. Math. Imaging Vis. **40**(1), 120–145 (2011)
37. C. Clason, T. Valkonen, Primal-dual extragradient methods for nonlinear nonsmooth PDE-constrained optimization. SIAM J. Optim. **27**(3), 1314–1339 (2017)
38. C. Lemarechal, Nondifferentiable optimisation subgradient and ε—subgradient methods, in *Optimization and Operations Research* (Springer, 1976), pp. 191–199
39. W. van Ackooij, R. Henrion, (sub-) gradient formulae for probability functions of random inequality systems under gaussian distribution. SIAM/ASA J. Uncertain. Quantif. **5**(1), 63–87 (2017)
40. A. Lekbir, H. Aissam, L. Amine, N. Mourad, A non-convex denoising model for impulse and Gaussian noise mixture removing using bi-level parameter identification. Inverse Probl. Imaging **16**(4), 827–870 (2022)
41. M. Nachaoui, L. Afraites, A. Laghrib, A regularization by denoising super-resolution method based on genetic algorithms. Signal Process. Image Commun. **99**, 116505 (2021)
42. L. Afraites, A. Hadri, A. Laghrib, M. Nachaoui, A high order PDE-constrained optimization for the image denoising problem. Inverse Probl. Sci. Eng. **29**(12), 1821–1863 (2021)
43. D.A. Murio, Implicit finite difference approximation for time fractional diffusion equations. Comput. Math. Appl. **56**(4), 1138–1145 (2008)
44. P. Zhuang, F. Liu, Implicit difference approximation for the time fractional diffusion equation. J. Appl. Math. Comput. **22**(3), 87–99 (2006)
45. A. Hadri, L. Afraites, A. Laghrib, M. Nachaoui, A novel image denoising approach based on a non-convex constrained PDE: application to ultrasound images. Signal Image Video Process. **15**, 1057–1064 (2021)
46. H. Badry, L. Oufni, H. Ouabi, H. Iwase, L. Afraites, A new fast algorithm to achieve the dose uniformity around high dose rate brachytherapy stepping source using Tikhonov regularization. Australas. Phys. Eng. Sci. Med. **42**, 757–769 (2019)

An Improvement to the Nonparametric Regression Models Using the Nonsmooth Loss Functions

Soufiane Lyaqini

Abstract This work addresses the supervised learning problem using a smooth and accurate approximation of the absolute loss function. The supervised learning problem is reformulated as an unconstrained minimization problem and solved using an accelerated and fast gradient descent method. Finally, we present some numerical results from real-life data sets to demonstrate the effectiveness of the proposed method and confirm that it is better in terms of stability and convergence speed compared to sub-gradient descent and Newton methods.

Keywords Supervised learning · Absolute loss function · Accelerated gradient descent method · Tikhonov regularization

1 Introduction

Machine learning is an artificial intelligence field that uses statistical, probabilistic, and optimization techniques to learn from previous examples and then utilize that past training data to classify new data [1]. Therefore, to create complicated models or algorithms in an automated manner, every learning process begins with determining a so-called learned model using a training data set [2]. The learned model is then tested against an independent training data set to assess how well it can generalize to unknown data. This is known as testing one. Machine learning has a wide range of applications that are both diverse and varied. Far from being comprehensive, we list a few well-known applications such as self-driving cars, speech recognition, medical diagnostics, biomonitoring, computation biology applications, interpretation of brain activity, as well as more [3–10].

In this paper, we are interested by the supervised learning algorithms. In this sense, we remark that there is a wide variety of supervised learning algorithms

S. Lyaqini (✉)
Ecole Nationale des Sciences Appliques, LAMSAD Laboratory, Hassan First University of Settat, Settat, Morocco
e-mail: lyaqini.soufiane@gmail.com

A. Laghrib et al. (eds.), *New Trends of Mathematical Inverse Problems and Applications*, Springer Proceedings in Mathematics & Statistics 428,
https://doi.org/10.1007/978-3-031-33069-8_2

available, each of which proposes a theoretical model in general that is formally defined as an optimization problem of minimizing error functionals over training set [2, 11–13]. Recently, considerable research has concentrated on establishing a link between supervised learning and the well-known inverse problems [14–18]. Furthermore, the supervised learning problem is linked to the inverse problem in the case of the L^1 loss function [19–22] based on a novel smoothing strategy that involves transforming the L^1 loss function into a differentiable and continuous function by employing a smoothing technique [23, 24]. Furthermore, the associate unconstrained minimization problem is directly solved by using Newton's method. Nevertheless, despite the quality of the obtained solution, this method still time consuming. In this work, we propose a fast and efficient algorithm based on accelerated gradient descent method for solving this supervised learning problem.

Because of the success of the absolute loss function in the modelization of various machine learning problems [19, 25, 26]. In this work, The supervised learning problem is reformulated to a regularized optimization one using the L^1 norm. The representer theorem allows us to reduce the optimization problem, which was originally presented in infinite dimensional spaces, to reproducing kernel Hilbert space. Kernel methods have proven to be effective on a wide range of real issues [27, 28]. Positive semidefinite kernels have been used successfully in a variety of disciplines. Furthermore, this tool effectively reduces computationally time-consuming tasks [29]. Then, to solve the resulting optimization problem, we opt for the accelerated gradient descent method, that has demonstrated its ability to deal against the non-smooth optimization problem [30, 31]. Finally, we present some numerical results from real-life data sets to demonstrate the efficiency of the proposed approach and confirm that it is better in terms of stability and convergence speed compared to sub-gradient descent and Newton.

The rest of this article Its structured as follows. In Sect. 2, we present the setting of the supervised learning problem and its transformation into a minimization one, using absolute loss function and Tikhonov regularization. The Sect. 3 is focused to theoretical tools on smooth approximation of absolute loss function. In Sect. 4 we present a numerical algorithm based on an accelerated gradient descent algorithm. In Sect. 5, we investigate the performance of the proposed algorithm by applying it to problems resulting from real-life experiences. the experiments demonstrate that the proposed algorithm is an effective and useful tool in machine learning problem.

2 Setting of the Problem

The objective of supervised learning, is to uncover an unknown function $f : \mathcal{X} \to \mathcal{Y}$ from training set $\left(x_j, y_j\right)_{j=1}^n$, where $\mathcal{X} \subset \mathbb{R}^d$ and $\mathcal{Y} \subset \mathbb{R}$. We do not expect $f(x_j) = y_j$ this because the random samples contain noise and other uncertainty, rather than that they are roughly equal. This approximation is typically achieved in supervised learning theory by minimizing some loss function. Given a training set

$\mathcal{T} = \left(x_j, y_j\right)_{j=1}^n$ and the hypothesis space $\mathcal{S}$ of all functions linking x and y. The issue is determining a model $f \in \mathcal{S}$ so that when $x \in \mathcal{X}$ is given, $f(x)$ predicts y. The best prediction function $f^\ddagger$ is defined as

$$f^\ddagger = arg \min_{f \in \mathcal{S}} \mathcal{D}(f(x_1), \ldots, f(x_n), y_1, \ldots, y_n), \tag{1}$$

where $\mathcal{D}$ is the gap between $f(x_j)$ and y_j for($i = 1, \ldots, n$).

It is preferable to utilize the reproducing kernel Hilbert space (RKHS) for the hypothesis space. For additional details, see [32].

Through the success off the L^1 loss function in several machine learning problems [19, 33], In the following section, we will look at

$$\mathcal{D}(f(x_1), \ldots, f(x_n), y_1, \ldots, y_n) = \|\mathrm{F}(x) - \mathrm{y}\|_1, \tag{2}$$

where $\mathrm{F}(x) = \begin{pmatrix} f(x_1) \\ \vdots \\ f(x_n) \end{pmatrix}$ and $y = \begin{pmatrix} y_1 \\ \vdots \\ y_n \end{pmatrix}$. Then the problem (1) get to be

$$\min_{\mathrm{F} \in \mathcal{S}^n} \|F(x) - \mathrm{y}\|_1, \tag{3}$$

which is in general ill-posed [34]. Then, we apply the Tikhonov regularization [35–37] in particular, which consists in replacing the minimization problem (3) with the problem.

$$\min_{\mathrm{F} \in \mathcal{S}^n} \left(\|\mathrm{F}(x) - \mathrm{y}\|_1 + \delta \|\mathrm{F}\|_{\mathcal{S}}^2\right), \tag{4}$$

where $\|.\|_{\mathcal{S}}$ is the norm in $\mathcal{S}$. and $\delta > 0$ is the regularization parameter.

Since $\mathcal{S}$ is infinite dimensional, It is not evident to solve the optimization problem (4). Because of the representer theorem [38], the problem (4) reduces to a finite dimensional optimization problem, as well as its solution is defined as

$$f(x) = \sum_{i=1}^n \lambda_i H(x, x_i), \quad with \quad \lambda_i \in \mathbb{R}. \tag{5}$$

where H is the reproducing kernel of $\mathcal{S}$. Then

$$F = H\lambda. \tag{6}$$

So the problem (4) can indeed be simplified to look for $\lambda = (\lambda_1, \ldots, \lambda_n)$ solution of

$$\min_{\lambda \in \mathbb{R}^n} \left(\left\| H\lambda - y \right\|_1 + \frac{\delta}{2} \|\lambda\|_H^2\right). \tag{7}$$

Defining

$$H = \begin{pmatrix} H(x_1, x_1) & \cdots & H(x_1, x_n) \\ \vdots & \ddots & \vdots \\ H(x_n, x_1) & \cdots & H(x_n, x_n) \end{pmatrix}, \; y = \begin{pmatrix} y_1 \\ \vdots \\ y_n \end{pmatrix} \text{ and } \lambda = \begin{pmatrix} \lambda_1 \\ \vdots \\ \lambda_n \end{pmatrix}.$$

Then the optimization problem is given by

$$\min_{\lambda \in \mathbb{R}^n} \left(\|H\lambda - y\|_1 + \frac{\delta}{2}\lambda^\top H \lambda \right). \tag{8}$$

3 Smoothed Absolute Loss

Using gradient descent methods to solve an optimization problem is a traditional method. So because L^1 norm in the optimization problem (8) is non-differentiable, global optimization methods can be used, such as sub-gradient method, which doesn't necessitate any condition of differentiability on the cost function. These techniques, however, are still time consuming [39, 40]. As a result, we use the so-called smooth approximation [41], in order to approximate the L^1 loss function with a continuous and differentiable one. This permits us to apply gradient methods kind to solve the ensuing optimization problem. Let us define $\mathcal{L}_\epsilon(u)$ the smoothed absolute function with parameter ϵ of $\mathcal{L}(u) = |u|$ defined by

$$\mathcal{L}_\epsilon(u) = \sqrt{u^2 + \epsilon^2},$$

where $\epsilon > 0$ is a smoothing parameter.

The function $\mathcal{L}_\epsilon$ is a uniform smooth approximation of absolute loss $\mathcal{L}$, in the following sense

Lemma 1 $\forall u \in \mathbb{R}$, *we have*

$$\mathcal{L}(u) \leq \mathcal{L}_\epsilon(u) \leq \mathcal{L}(u) + \epsilon. \tag{9}$$

Proof This in fact flows from the fact that

$$u^2 \leq u^2 + \epsilon^2 \leq (|u| + \epsilon)^2.$$

So by applying the square function, we get

$$|u| \leq \sqrt{u^2 + \epsilon^2} \leq |u| + \epsilon.$$

Then

$$\mathcal{L}(u) \leq \mathcal{L}_\epsilon(u) \leq \mathcal{L}(u) + \epsilon.$$

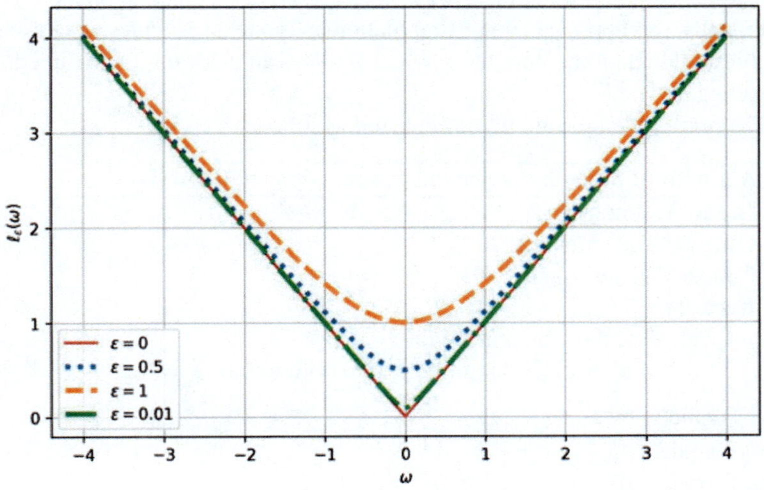

Fig. 1 Smoothed absolute loss function with different smoothing parameters

According to this Lemma, we have that $\mathcal{L}_\epsilon$ converges to $\mathcal{L}$ when ϵ go to zero. This is illustrated in fact in the flowing figure (Fig. 1).

Concurrently, this approach aims to replace $\|H\lambda - y\|_1$ by the following approximate quadratic formula

$$\frac{1}{n}\sum_{i=1}^{n}\left(\left(\lambda^T H_i - y_i\right)^2 + \epsilon^2\right)^{\frac{1}{2}}, \qquad \text{for } \epsilon > 0, \tag{10}$$

which is convex and differentiable. Then the optimization problem (8) become

$$\min_{\lambda\in\mathbb{R}^n}\left(\frac{1}{n}\sum_{i=1}^{n}\left(\left(\lambda^T H_i - y_i\right)^2 + \epsilon^2\right)^{\frac{1}{2}} + \frac{\delta}{2}\lambda^\top H\lambda\right), \qquad \text{for } \epsilon > 0, \tag{11}$$

where $H_i = \left(H(x_i, x_j)\right)_{j=1}^{n}$.

4 Numerical Algorithms

There are various optimization methods available to solve the problem (11) (See for example [42, 43]). to solve the problem (11) we use the acceleration methods with line search methods.

The accelerated gradient descent method is well-known for achieving the best convergence rate for first order approaches on convex problems [43, 44]. It deals, in particular, in improving the the convergence rate of gradient descent methods. more specifically, this approach is an extrapolation method that provides weight sequences for the so-called momentum term, which updates an iteration in the direction of the preceding update.

The proposed algorithm is summarized as follows:

Algorithm: Smooth accelerated gradient descent **(S-AGD)**

Input: Training set Z

Output: The parameter λ^{ϵ}

Choose β_0^{ϵ} ,ϵ, σ, t and δ

Iteration:

1 Calculate the Gradient G_t^{ϵ} on $\mathcal{T}$

2 Find γ_t by backtracking line search with initial $\gamma_0 = 0.2$.

3 Find τ_t by $s_0 = 0, \quad s_t = \frac{1+\sqrt{1+4s_{t-1}^2}}{2}$, $\tau_t = \frac{s_{t-1}-1}{s_t}$

4 Update $\beta_{t+1}^{\epsilon} = \alpha_t^{\epsilon} - \gamma_t.G_k^{\epsilon}$

5 Update $\alpha_{t+1}^{\epsilon} = \beta_{t+1}^{\epsilon} + \tau_t(\beta_{t+1}^{\epsilon} - \beta_t^{\epsilon})$

5 Experiments

In this section, we evaluated our algorithms, we selects *Airfoil Self-Noise (ASN) and Combined Cycle Power Plant (CCPP)* UCI datasets [45] for experimentation. Tables 1, 2, 3 and Fig. 2 depicts the description and the heatmap of the correlation matrix, respectively, of each dataset. Furthermore, to investigate the effectiveness of the proposed we compared its computational time and its relative error with those of the Sub-gradient descent **(Sub-GD)** and smooth Newton **(S-Newton)** algorithms. The goal is to demonstrate that the proposed approach accurately and quickly predicts the models under consideration.

Table 1 Data sets

Data sets	Number of instances	Number of attributes
ASN	1503	6
CCPP	9568	4

Table 2 Description of CCPP dataset

Meaning	Name	Range
Ambient temperature	AT	1.81–37.11°C
Ambient pressure	AP	992.89–1033.30 millibar
Relative humidity	RH	25.56%–100.16%
Exhaust vacuum	V	25.36–81.56 cm Hg
Net hourly electrical energy output	PE	420.26–405.76 MW

Table 3 Description of ASN dataset

Meaning	Name	Range
Frequency	f	in Hertzs
Angle of attack	alpha	in degrees
Chord length	c	in meters
Free-stream velocity	U-infinity	in meters per second
Suction side displacement thickness	delta	in meters
Scaled sound pressure level	SSPL	in decibels

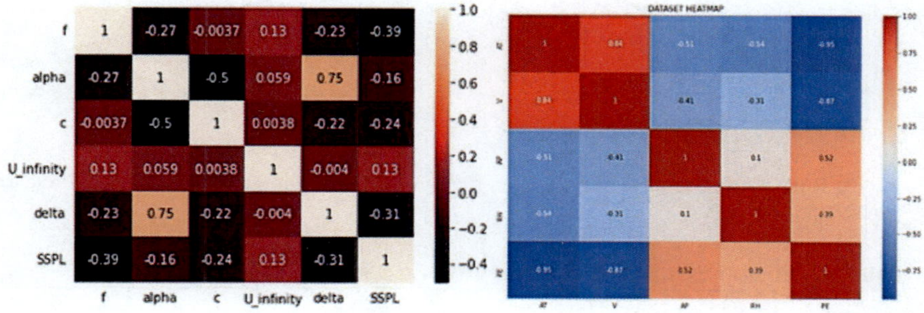

(a) (b)

Fig. 2 Heatmap of Correlation matrix. **a** ASN and **b** CCPP

The kernel function used in the experiments is $H(x, y) = \exp\left(-\theta\|x - y\|^2\right)$, where θ is a parameter in the kernel function. In addition, for all the experiments we set $\delta = 5.10^{-5}$ and $\epsilon = 10^{-3}$. All data points are normalized to [0, 1] according to the minimum and maximum values in the corresponding feature column.

Figure 3 and Table 4 respectively, show the results of the running time, cost function and the relative error for the **S-AGD**, **SNewton** and **Sub-GD** algorithms. The results showed that the **S-AGD** algorithm is better in terms of the running speed and second best relative error. This results demonstrate that the proposed algorithm achieve the optimal solution with a good accuracy, but without sacrificing training speed.

5.1 Stability of the Algorithm

To further provide the anti-noise behavior of the proposed algorithm, the real-world dataset Combined Cycle Power Plant (with noise) from UCI was employed in the experiment. So, Gaussian noise is added to each feature with a mean of 0 and variance of 1. The experiment is done in tow cases, which are: $r = 5\%$ and $r = 10\%$ of levels noise. Figure 4 demonstrates the prediction model closely mimics the real one's behavior, which indicating that the proposed algorithm is stable. This demonstrates that the suggested method is a regularizing one.

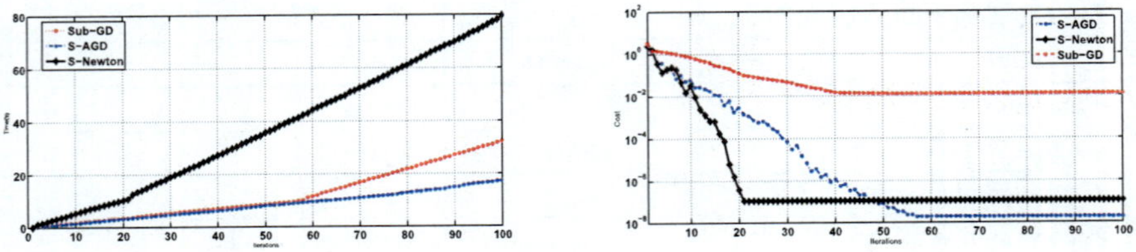

(a)

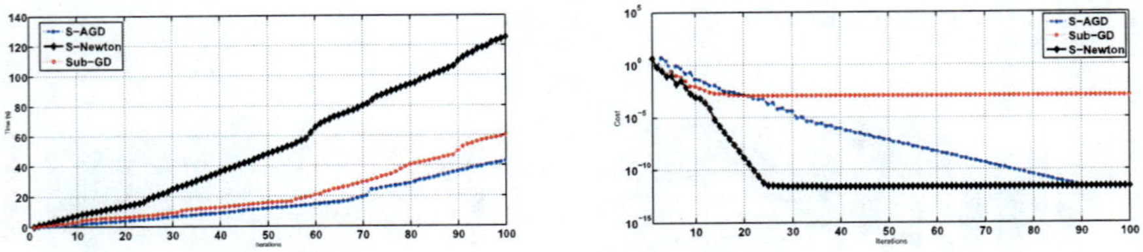

(b)

Fig. 3 Running time and the cost function for **S-AGD S-Newton** and **Sub-GD** on UCI datasets. **a** ASN and **b** CCPP

Table 4 The relative error for **S-AGD S-Newton** and **Sub-GD** on UCI datasets

	S-AGD	S-Newton	Sub-GD
ASN	0.0022	0.0017	0.0083
CCPP	0.027	0.021	0.091

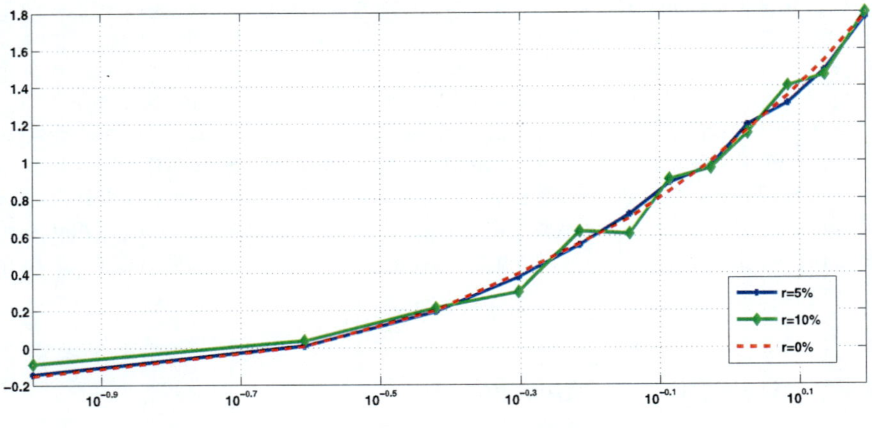

Fig. 4 Perturbed model for **S-AGD** algorithm on Airfoil Self-Noise dataset

6 Conclusion

In this paper, in order to solve the supervised learning problem, we proposed the smoothed accelerated gradient descent **(S-AGD)** algorithm based on nonsmooth loss function. Experimental results and statistical tests showed that the **(S-AGD)** algorithm is the fastest method when compared to other algorithms, and it also ensures extremely excellent accuracy and stable solutions. Therefore, we will consider how to extend the proposed algorithm in classification and time-series problems

References

1. V.N. Vapnik, *The Nature of Statistical Learning Theory* (1999)
2. V.N. Vapnik, *The Nature of Statistical Learning Theory* (Springer, Berlin, Heidelberg, 1995)
3. H. Chen, R. Chiang, V. Storey, Business intelligence and analytics: from big data to big impact. MIS Q.: Manag. Inf. Syst. **36**, 1165–1188 (2012)
4. L. Devroye, L. Györfi, G. Lugosi, A probabilistic theory of pattern recognition, in *Applications of Mathematics*, vol. 31 (Springer, New York, 1996). https://doi.org/10.1007/978-1-4612-0711-5
5. J. Rozas, J.C. Sánchez-DelBarrio, X. Messeguer, R. Rozas, Dnasp, DNA polymorphism analyses by the coalescent and other methods 19, 2496–7 (2004)
6. P. Rani, C. Liu, N. Sarkar, E. Vanman, An empirical study of machine learning techniques for affect recognition in human robot interaction. Pattern Anal. Appl. **9**, 58–69 (2006)
7. I. Kononenko, Machine learning for medical diagnosis: history, state of the art and perspective. Artif. Intell. Med. **23**, 89–109 (2001)
8. E. Osuna, R. Freund, F. Girosi, Training support vector machines: an application to face detection, 130–136 (1997)
9. M. Nachaoui, A. Laghrib, An improved bilevel optimization approach for image super-resolution based on a fractional diffusion tensor. J. Frankl. Inst. **359**(13), 7165–7195 (2022)
10. D. Kumar, U. Bansal, R. Alroobaea, A. Baqasah, M. Hedabou, An artificial intelligence approach for expurgating edible and non-edible items. Front. Public Health **9** (2022). https://doi.org/10.3389/fpubh.2021.825468
11. F. Cucker, S. Smale, On the mathematical foundations of learning. Bull. Am. Math. Soc. **39**, 1–49 (2002)
12. K. Slavakis, G. Giannakis, G. Mateos, Modeling and optimization for big data analytics: (statistical) learning tools for our era of data deluge, 31, 18–31 (2014)
13. A. Emrouznejad, *Big Data Optimization: Recent Developments and Challenges*, vol. 18 (2016)
14. C.D.M.M. Bertero, E.R. Pike, Linear inverse problems with discrete data: Ii-stability and regularization. Inverse Probl. **4**, 573–594 (1988)
15. A. Kirsch, *An Introduction to the Mathematical Theory of Inverse Problems* (Springer, Berlin, Heidelberg, 1996)
16. V. Kůrková, Supervised learning as an inverse problem, 1377–1384 (2004)
17. S. Mukherjee, P. Niyogi, T. Poggio, R. Rifkin, Learning theory: stability is sufficient for generalization and necessary and sufficient for consistency of empirical risk minimization. Adv. Comput. Math. **25**, 161–193 (2006)
18. E. De Vito, L. Rosasco, A. Caponnetto, M. Piana, A. Verri, Some properties of regularized kernel methods. J. Mach. Learn. Res. **5**, 1363–1390 (2004)
19. S. Lyaqini, M. Quafafou, M. Nachaoui, A. Chakib, Supervised learning as an inverse problem based on non-smooth loss function. Knowl. Inf. Syst. 1–20 (2020)

20. S. Lyaqini, M. Nachaoui, Identification of genuine from fake banknotes using an enhanced machine learning approach, in *International Conference on Numerical Analysis and Optimization Days* (Springer, 2021), pp. 59–70
21. S. Lyaqini, M. Nachaoui, A. Hadri, An efficient primal-dual method for solving non-smooth machine learning problem. Chaos, Solitons & Fractals **155**, 111754 (2022)
22. A. Hadri, M. Nachaoui, A. Laghrib, A. Chakib, L. Afraites, A primal-dual approach for the robin inverse problem in a nonlinear elliptic equation: the case of the *l1 l2* cost functional. J. Inverse Ill-Posed Probl. (2022) [cited 2022-10-23]. https://doi.org/10.1515/jiip-2019-0098
23. Y.-J. Lee, W.-F. Hsieh, C.-M. Huang, Epsilon-SSVR: a smooth support vector machine for epsilon-insensitive regression. IEEE Trans. Knowl. Data Eng. (5), 678–685 (2005)
24. J. Hajewski, S. Oliveira, D. Stewart, Smoothed hinge loss and ℓ^1 support vector machines, in *2018 IEEE International Conference on Data Mining Workshops (ICDMW)* (IEEE, 2018), pp. 1217–1223
25. S. Lyaqini, M. Nachaoui, M. Quafafou, Non-smooth classification model based on new smoothing technique. J. Phys.: Conf. Ser. **1743**(1), 012025 (2021)
26. S. Lyaqini, M. Nachaoui, Diabetes prediction using an improved machine learning approach. Math. Model. Comput. **8**, 726–735 (2021). https://doi.org/10.23939/mmc2021.04.726
27. P. Ponte, R.G. Melko, Kernel methods for interpretable machine learning of order parameters. Phys. Rev. B **96**(20), 205146 (2017)
28. A. Apsemidis, S. Psarakis, J.M. Moguerza, A review of machine learning kernel methods in statistical process monitoring. Comput. Ind. Eng. **142**, 106376 (2020)
29. T. Hofmann, B. Schölkopf, A.J. Smola, Kernel methods in machine learning. Ann. Stat. 1171–1220 (2008)
30. D. Mitchell, N. Ye, H. De Sterck, Nesterov acceleration of alternating least squares for canonical tensor decomposition (2018), arXiv:1810.05846
31. M. Nachaoui, L. Afraites, A. Laghrib, A regularization by denoising super-resolution method based on genetic algorithms. Signal Process.: Image Commun. **99**, 116505 (2021)
32. N. Aronszajn, Theory of reproducing kernels. Trans. Am. Math. Soc. **68**, 337–404 (1950)
33. L. Rosasco, E. De Vito, A. Caponnetto, M. Piana, A. Verri, Are loss functions all the same? Neural Comput. **16**, 1063–1076 (2004)
34. J. Hadamard, *Lectures on Cauchy's Problem in Linear Partial Differential Equations* (Yale University Press, New Haven, 1923)
35. M. Nachaoui, L. Afraites, A. Hadri, A. Laghrib, A non-convex non-smooth bi-level parameter learning for impulse and Gaussian noise mixture removing. Commun. Pure Appl. Anal. **21**(4), 1249 (2022)
36. L. Afraites, A. Hadri, A. Laghrib, M. Nachaoui, A non-convex denoising model for impulse and gaussian noise mixture removing using bi-level parameter identification. Inverse Probl. Imaging (2022)
37. L. Afraites, A. Atlas, Parameters identification in the mathematical model of immune competition cells. J. Inverse Ill-Posed Probl. **23**(4), 323–337 (2015)
38. B. Schölkopf, R. Herbrich, A.J. Smola, A generalized representer theorem, in *International Conference on Computational Learning Theory* (Springer, 2001), pp. 416–426
39. C. Lemarechal, Nondifferentiable optimisation subgradient and ε-subgradient methods, in *Optimization and Operations Research* (Springer, 1976), pp. 191–199
40. W. van Ackooij, R. Henrion, (Sub-) Gradient formulae for probability functions of random inequality systems under Gaussian distribution. SIAM/ASA J. Uncertain. Quantif. **5**(1), 63–87 (2017)
41. R. Rosales, M. Schmidt, G. Fung, Fast optimization methods for l1 regularization: a comparative study and two new approaches (2007)
42. S. Boyd, L. Vandenberghe, *Convex Optimization* (Cambridge University Press, New York, 2004)
43. S. Ruder, An overview of gradient descent optimization algorithms (2016), arXiv:1609.04747
44. B. O'donoghue, E. Candes, Adaptive restart for accelerated gradient schemes. Found. Comput. Math. **15**(3), 715–732 (2015)
45. https://archive.ics.uci.edu/ml/datasets/airfoil+self-noise

The Asymptotic Behavior of the Reynolds Equations' Solution

Youssef Essadaoui and Imad Hafidi

Abstract The work deals with the dynamic analysis of a mechanical system consisting of two rigid surfaces, one of which is in motion relative to the other. We study the asymptotic behavior of the solution to the compressible and incompressible Reynolds equations, also for the Eldord and Adams models, in the case of a single degree of freedom.

Keywords Reynolds equations · Asymptotic behavior · Fluid interaction

1 Introduction

In all the mathematical work relating to lubrication, the problem is to study existence, the uniqueness, qualitative or quantitative properties of the possible solution of a Reynolds-type equation where data are related to the geometry or rheology of the fluid and the unknown is pressure.

From this pressure it is possible to obtain by simple integration the overall operational characteristics of the lubricated mechanism such as the load borne by it or the moment.

In reality, in most real problems, only external forces acting on the mechanical system are known. The latter adapts to meet these external demands and it is therefore important to know whether the corresponding equilibrium points are acceptable from the point of view of operational safety [minimum thickness between surfaces to avoid contact, stability of equilibrium position]. Very few mathematical studies have been devoted to this topic [1, 2].

Y. Essadaoui (✉) · I. Hafidi
National School of Applied Sciences, Sultan Moulay Slimane University, Bd Beni Amir, BP 77, Khouribga, Morocco
e-mail: youssefessadaoui2017@gmail.com

A. Laghrib et al. (eds.), *New Trends of Mathematical Inverse Problems and Applications*, Springer Proceedings in Mathematics & Statistics 428,
https://doi.org/10.1007/978-3-031-33069-8_3

We consider as model problem a device of the type "Articulated Slider" two degrees of freedom. The first is vertical displacement, a, and the second is the angle of rotation θ around an axis located at the x-axis x_0 The external force that this mechanism must support is a vertical force $F(t)$ (parallel to Ox_3). The hydrodynamic pressure $p(x,t)$ induces a lift $W(t)$ given by:

$$W(t) = \int_{\Omega} p(x,t)\mathrm{dx}$$

In practice, we want it to balance the force $F(t)$. The problem is the "reciprocal" version of the problem, which is usually studied in many mathematical works in lubrication. These works attempt to obtain existence and uniqueness results for equation Reynolds for a given geometry h (see, for example, [3–5]).

We consider the following static problem:

$$\begin{cases} \nabla \cdot \left[h^3(x)\rho\nabla p\right] = \partial_{x_1}(\rho h) \quad x = (x_1, x_2) \in \Omega \\ p(x,t) = p_a \geq 0 \quad x \in \partial\Omega \\ \int_{\Omega} p(x)\mathrm{dx} = F. \end{cases} \tag{1}$$

ρ the density.

The space between the two surfaces is filled with incompressible or compressible lubricating fluid in the context of thin film mechanics context of thin film mechanics, this fluid obeys the Reynolds equation. The first part of this work deals with an incompressible dynamic problem.

2 The Incompressible Case

We model the fluid pressure by the incompressible Reynolds equation, the variational inequation, and the Eldrod-Adams model.

2.1 Incompressible Reynolds Equation

The Reynolds equation for incompressible fluids is an equation obtained rigorously from the Navier-Stokes equations when the domain thickness tends to 0. This theoretical result has been justified for more than twenty years [3].

In this part we consider the case of a single degree of freedom. We model the fluid pressures by the incompressible Reynolds equation. The formulation of our problem is the following:

Find $p(x)$ and $\epsilon > 0$ solutions of

$$\begin{cases} \nabla \cdot \left[\left(\hat{h}_0 + \epsilon \right)^3 \nabla p \right] = \frac{\partial \hat{h}_0}{\partial x_1} & x \in \Omega \\ p = 0 & \text{on } \partial\Omega \\ \int_\Omega p = F & \\ \hat{h}_0 \geq 0 \text{ fixed and } \min_{x\in\bar{\Omega}} \hat{h}_0(x) = 0 & \end{cases} \tag{2}$$

In the one-dimensional case and for $\hat{h}_0$ affine(flat upper surface) the problem has been solved analytically in [6]. As intuitively expected the author showed that the equilibrium distance a decreases as the applied force F increases. In the case where $\hat{h}_0$ is no longer affine or in the two dimensional case the only results that exist are purely numerical [7].

The solution method proposed for the problem in [6] is based on the introduction of the:

$$F = \int_\Omega p(x)\mathrm{dx}$$

where p is the solution of problem (2).

The moments of the force exerted by the pressure for each minimal gap thickness ϵ defined with respect to the point x^0,

$$m_i(\epsilon) = \int_\Omega p\left(x_i - x_i^0\right) dx \quad i = 1, \ldots, n \tag{3}$$

because equilibrium also requires that

$$m_i(\epsilon) = 0 \quad i = 1, \ldots, n$$

In the paper [8], they assume $n = 2, \Omega =]a_1, b_1[\times]a_2, b_2\left[, h_0 \in C^0(\bar{\Omega})\right.$ with $h_0(x) > 0$ a.e $x \in \Omega$, and

$$\min_{x\in\bar{\Omega}} h_0(x) = 0$$

Two particular geometrical configurations have been studied, each corresponding to specific mechanisms.

1- Line contact: $\hat{h}_0$ cancels on a segment $\{x_1 = 0\}$ this case can appear in journal bearing. We consider that $\hat{h}_0$ behaves like $(-x_1)^\alpha$.

2- Point contact [point contact]: $\hat{h}_0$ cancels at the point (0, 0) which can occur in ball bearings. Consider $\hat{h}_0$ behaves as $\left(x_1^2 + x_2^2\right)^{\alpha/2} = |x|^\alpha$.

Finite limit case

Theorem 1 *For all* $\alpha \in]0, 1[$ *and* $m_0 > 0$ *such that:*

$$h_0(x) \geq m_0\,(x_1)^\alpha \quad x \in \Omega$$

Then for all $(x_1^0, x_2^0) \in \Omega$*, we have* $\epsilon \to 0$*:*

$$\int_\Omega p dx \to \int_\Omega p_0 dx, \quad \int_\Omega (x_k - x_k^0)\, p \to \int_\Omega (x_k - x_k^0)\, p_0, k = 1, 2$$

such that p *is a solution of problem (2) and* p_0 *the solution of problem (5).*

We can now give the main result in the case $\alpha \geq 1$.

Infinite limit case

Theorem 2 *For* $\alpha \geq 1$ *we have* $\int_\Omega p dx \to +\infty$ *as* $\epsilon \to 0$*. Moreover there exists* $K > 0$ *such that for* ϵ *small enough we have*

$$\int_\Omega p dx \geq K \epsilon^{\frac{2}{\alpha} - 2} \quad \text{for } \alpha > 1$$

$$\int_\Omega p dx \geq K \log\left(\frac{1}{\epsilon}\right) \quad \text{for } \alpha = 1$$

2.2 Variational Inequation

The problem is considered in the case of variational inequation. The solution of the problem (2) can then take negative values [9]. It is then usual to replace the problem (2) by the associated variational inequation. The problem to be solved is then:

Find $(p, a) \in K \times \mathbb{R} \times \mathbb{R}$ such as

$$\begin{aligned} &\int_\Omega (h_0 + a)^3\, p' \left(\varphi' - p'\right) \geq \int_\Omega (h_0 + a) \left(\varphi' - p'\right) \quad \forall \varphi \in K \\ &\int_\Omega p = F. \end{aligned} \tag{4}$$

where $K = \left\{\varphi \in H_0^1(\Omega), \varphi \geq 0\right\}$. We obtain the same results using the results of the Reynolds equation [8].

2.3 Model Eldord and Adams

In this last section, we study the equilibrium problem with a single degree of freedom by modeling the fluid pressure with the Eldrod and Adams model. The formulation of our problem is as follows:

Find $p(x)$ and $a > 0$ solutions of

$$\begin{array}{lll} \nabla \cdot \left[\left(\hat{h}_0 + a \right)^3 \nabla p \right] = \frac{\partial \hat{h}_0}{\partial x_1} & x \in \Omega \\ \int_\Omega p = F \quad p = 0 & on\ \partial\Omega \end{array} \tag{5}$$

To simplify, we take: $\Omega =]0,1[\times]0,1[$. h_ϵ verifies the following assumptions:

$$h_\epsilon = h_0 + \epsilon \quad h_0 \in L^\infty(\Omega) \quad \epsilon > 0 \quad \min_{x \in \bar{\Omega}} h_0 = 0$$

Following the same approach as for the Reynolds equation, we study the asymptotic behavior of the solution of the problem (5) when $\epsilon \to 0$ in order to solve our problem. Let q be the solution of the Reynolds equation:

$$\begin{cases} \int_{\tilde{\Omega}} h^3 \nabla q \nabla \varphi == \int_{\tilde{\Omega}} h \frac{\partial \varphi}{\partial x_1} \\ V = \left\{ \varphi \in H_0^1(\tilde{\Omega}) \text{ such that } \varphi = 0 x \in \partial\tilde{\Omega} \right\} \end{cases} \quad \forall \varphi \in V$$

where:

$$\tilde{\Omega} \subseteq \left\{ x \in \Omega \text{ such that } \frac{\partial h_0}{\partial x_1} \leqslant 0 \right\}$$

and

$$\{x \in \omega \text{ such that } h \equiv 0\} \bigcap \tilde{\Omega} \neq \emptyset$$

We will use the maximum principle to prove that $p_\epsilon \geqslant q_\epsilon$ on $\tilde{\Omega}$. We study the one dimensional case, and we use analytical calculation in dimension 1 to locate a part of the saturated zone which can include $\tilde{\Omega}$, the principle of the maximum is a natural result [9].

Theorem 3 *Let h_0 be in the form:*

$$h_0 = (x_1 - a)^\alpha h_1$$

such that $h_1 \neq 0$ on $\tilde{\Omega}$ and $\alpha > 1$. The problem (5) has a solution.

3 The Compressible Case

In the case of compressible fluids, this equation is very much used with fluids that are compressible. These equations are often used, for example, in problems related to magnetic hard disk drives and magnetic tapes, a very thin film of air is located between the read head and the media. The mathematical justification of this compressible Reynolds equation has been proved only very recently (see the works of E. Marusic-

Paloka and M. Starcevic [10, 11]), and only in the case of perfect gases. One of the main reasons for this lack of mathematical results comes, undoubtedly, from the intrinsic difficulty linked to the compressible Navier-Stokes equations.

3.1 Reynolds Compressible

We will consider in this section the case where the lubricating fluid between two rigid surfaces is a compressible and isothermal fluid (air, for example). The normalized pressure of the fluid verifies:

$$\begin{aligned} &\nabla\cdot\left[\left(H^3p+\lambda H^2\right)\nabla p\right]=\frac{\partial}{\partial x_1}[Hp] \quad x=(x_1,x_2)\in\Omega \\ &p=1 \quad x\in\partial\Omega \end{aligned} \tag{6}$$

where H is the normalized distance between the surfaces with $H(x)=h_0(x)+\epsilon$ with $\epsilon>0$, Ω is an open bounded of $\mathbb{R}^n$, $n=1,2$, $\lambda\in[0,+\infty[$ is the rarefaction parameter of the air.

For simplicity, we consider $\Omega=]-1,0[^2$. Throughout this paragraph, it is assumed that h_0 satisfies the following assumptions:

$$(H)\begin{cases} h_0\in L^\infty(\Omega) \text{ and we note } M_0=\|h_0\|_{L^\infty(\Omega)} \\ \inf_{\bar{\Omega}} h_0(x)=0, \qquad h_0>0 \text{ on } \Omega \\ h_0 \text{ is decreasing with respect to } \quad x_1\in\Omega \end{cases} \tag{7}$$

The goal is to adapt the techniques used in Sect. 2.1 to study the asymptotic behavior of the charge and moments when $\epsilon\to 0$. The asymptotic behavior is going to be different depending on the assumptions checked by h_0. First, we will prove that the starting problem with ϵ fixed (6) is well posed. We will introduce, for this purpose, Sobolev spaces with weights and we will show an inequality of Poincaré type which will be used to study the convergence of the solution of (6).

Under the hypothesis $\frac{1}{h_0}\in L^1(\Omega)$ we define the limit problem in two different ways. To begin, we define a first problem with test functions of type $\mathcal{D}(\Omega)$. This type of test function does not allow us to have the uniqueness of the limit problem. We will then be led to define a new limit problem whose test functions belong to a more restricted space. This leads to an additional hypothesis on h_0. Although the techniques of proof are almost identical, we will be obliged in this case to differentiate between the case $\lambda=0$ and the case $\lambda\neq 0$. As an immediate result, we obtain the convergence of the load and moments in the same framework. The case where the condition $\frac{1}{h_0}\in L^1(\Omega)$ is not verified, the limit problems are not well posed. The non-convergence of the load and moments when $\epsilon\to 0$ is proved by using the maximum principle.

We consider the case of a single degree of freedom for the upper surface, we will show the existence of a solution for the following equilibrium problem: Find $(p, \epsilon) \in \left(1 + H_0^1(\Omega)\right) \times \mathbb{R}^+$ such as:

$$\begin{cases} \nabla \cdot \left[\left[(h_0 + \epsilon)^3\, p + \lambda\, (h_0 + \epsilon)^2\right] \nabla p\right] = \frac{\partial (h_0+\epsilon) p}{\partial x_1} & \text{in } \Omega \\ p = 1 & \text{on } \partial\Omega \\ \int_\Omega (p-1) \mathrm{dx} = F \end{cases} \tag{8}$$

where ϵ is the vertical displacement of the upper surface.

3.1.1 Preliminary Results

Finite limit case

In this section, in addition to (8), we will assume that h_0 satisfies

$$(H_0) \quad \begin{cases} \frac{1}{h_0} \in L^1(\Omega) \\ K_0 = \sup_{x_2 \in [-1,0]} \int_{-1}^{0} \int_{-1}^{x_1} \frac{1}{h_0^2} \mathrm{dsdx}_1 < +\infty. \end{cases}$$

We study the asymptotic behavior of the weak solution of the problem (6) when ϵ tends to 0 . We introduce the following limit problem:

$$\text{Find } p_0 \in 1 + H_0^1\left(\Omega, h_0, h_0^3\right) \text{ such that}$$

$$\int_\Omega \left(h_0^3 p_0 + \lambda h_0^2\right) \nabla p_0 \nabla \varphi \mathrm{dx} = \int_\Omega h_0 p_0 \frac{\partial \varphi}{\partial x_1} \mathrm{dx} \quad \forall \varphi \in \mathcal{D}(\Omega). \tag{9}$$

We need the weighted Sobolev space $H^1\left(\Omega, H, H^3\right)$ with $H(x) = h_0(x) + \epsilon$. We can now state the result principle:

Theorem 4 *Under assumption* (H_0)*, a solution* p_0 *exists for* $\frac{1}{h_0^3} \in L^1(\Omega)$*, such that for all* $\left(x_1^0, x_2^0\right) \in \Omega$ *we obtain, up to a subsequence of* ϵ*,*

$$\int_\Omega p\mathrm{dx} \to \int_\Omega p_0 \mathrm{dx}, \int_\Omega \left(x_k - x_k^0\right) p\mathrm{dx} \to \int_\Omega \left(x_k - x_k^0\right) p_0 \mathrm{dx}, \quad k = 1, 2. \tag{10}$$

If, in addition, $\frac{1}{h_0^3} \in L^1(\Omega)$*, then for all sequences of* ϵ*, we have the same convergences (problem (9)), with* p_0 *as the unique solution of (problem 3.15, [12]) for* $\lambda = 0$ *or of (problem 3.16, [12]) in the case of* $\lambda > 0$*.*

Infinite-limit case

The results obtained in the preceding section apply in particular for the case $h_0 \sim (-x_1)^\alpha$ in a neighborhood of $\{x_1 = 0\}$, with $0 < \alpha < 1$. We consider in this section the case $\alpha > 1$ for $\lambda = 0$ or $\alpha \geq 2$ for $\lambda > 0$. Since the limit problems (9), (problem 3.15, and 3.16, [12]) are not valid in this case, we will use the maximum principle to study the asymptotic behaviour for the p solution of (6) when $\epsilon \to 0$ In this section we need the following hypothesis.

$$(H_1)\begin{cases}\text{There exist } \alpha > 1 \text{ and } h_1 \in C^1(\bar{\Omega}) \text{ with } h_1 > 0 \text{ such that}\\ h_0(x) = (-x_1)^{\alpha}\, h_1(x), \quad \forall x \in \bar{\Omega}\\ M_0 < \alpha h_1(0, x_2), \quad \text{for all } x_2 \in [-1, 0] \quad \text{with } M = \sup_{x\in\Omega} h_0(x).\end{cases}$$

$$(H_2)\begin{cases}\text{There exist } 0 < k_1 < k_2 \text{ such that}\\ k_1 (-x_1)^{\alpha-1} \le -\frac{\partial h_0}{\partial x_1} \le k_2 (-x_1)^{\alpha-1}\end{cases} \quad \forall x \in \Omega.$$

We introduce the following notation.

$$m_0 = \inf_{x\in\bar{\Omega}} h_1(x), \quad M_i = \sup_{x\in\Omega}\left|\frac{\partial h_1(x)}{\partial x_i}\right|, i = 1, 2, \quad A = \sup_{x_2\in[-1,0]} \frac{M_0}{\alpha h_1(0, x_2)} < 1$$

Let p_1 and p_2 be two positive functions defined by

$$p_1(x_1) = \frac{(-x_1)^{3\alpha-1}}{(M_0 (-x_1)^{\alpha} + \epsilon)^3} \quad \forall x_1 \in [-1, 0]$$

$$p_2(x_2) = x_2^2 (1 + x_2)^2 \quad \forall x_2 \in [-1, 0]$$

We will use, in what follows, the maximum principle ([12], Lemma 2.10), to increase p by the function $cp_1p_2 + c^{1/2}$ with $c > 0$ as a small enough constant independent of ϵ. This will enable us to conclude the asymptotic behavior of p.

Theorem 5 *Assume that $\alpha > 1$ for $\lambda = 0$ and $\alpha \ge 2$ for $\lambda > 0$; we then have, for all $\delta \in (0, 1)$,*

$$\int_{\Omega_\delta} p dx \to +\infty \quad \text{for } \epsilon \to 0$$

with $\Omega_\delta = (-\delta, 0) \times (-1, 0)$. In particular we then have

$$\int_{\Omega} p dx \to +\infty \quad \text{for } \epsilon \to 0.$$

Moreover, $R > 0$ exists as a constant depending of δ, such that for a small enough δ we then have

$$\int_{\Omega_\delta} p\mathrm{dx} \ge R \log\left(\frac{1}{\epsilon}\right)$$

In the following [8], we will give results on the asymptotic behavior of moments.

Theorem 6 *Under the assumptions of Theorem 6, if moreover h_0 is symmetric in x_2 with respect to $x_2 = -1/2$, we then have, for $\epsilon \to 0$,*
- $\int_\Omega \left(x_2 - x_2^0\right) p\mathrm{dx} \to +\infty$ *if $x_2^0 \in \left(-1, -\frac{1}{2}\right)$;*

- $\int_{\Omega}\left(x_2 - x_2^0\right) p\mathrm{dx} \to -\infty$ *if* $x_2^0 \in \left(-\frac{1}{2}, 0\right)$;
- $\int_{\Omega}\left(x_2 - x_2^0\right) p\mathrm{dx} \to 0$ *if* $x_2^0 = -\frac{1}{2}$.

We have the following behavior of the load for $\lambda = 0$.

Theorem 7 *Under the assumptions of the previous lemma and Theorem 6, we obtain*

$$\int_{\Omega}\left(x_1 - x_1^0\right) p\mathrm{dx} \to +\infty \text{ if } \epsilon \to 0 \text{ for all } x_1^0 \in (-1, 0). \tag{11}$$

References

1. J.I. Diaz et, J. Ignacio Tello, On an inverse problem arising lubrication theory differential integral equations **17**(5), 583–591 (2004)
2. V.V. Jikov, S.M. Kozlov, O.A. Oleinik, *Homogenization of Differential Operators and Integral Functionals* (Springer, Berlin, 1994), p. 570
3. G. Bayada, M. Chambat, The transition between the Stokes equations and the Reynolds equation: a mathematical proof. Appl. Math. Optim. **14**(1), 73–93 (1986)
4. C.C. Vazquez, Existence and uniqueness of solution for a lubrication problems with cavitation in a journal bearing with axial supply. Adv. Math. Sci. Appl. **42**, 313–331 (1994)
5. K. Taous, Equations de Reynolds pour une large clase de fluides non-Newtoniens. C. R. Acad. Sci. Paris Sé. I Math. **323**(11), 1213–1218 (1996)
6. L. Leloup, *Etude de la lubrification et calcul des paliers: lois theoriques et experimentales*, 2e edn. (Dunod, Paris, 1962), p. 335
7. J. Frene, D. Nicolas, B. Degueurce, D. Berthe et, M. Godet, Lubrification hydrody- namique. Paliers et butées, EYROLLES, p. 488 (1990)
8. I. Ciuperca, I. Hafidi, M. Jai, Singular perturbation problem for the incompressible Reynolds equation, Submitted. Also available as part of Hafidi's thesis (Thése de Doctorat, INSA de Lyon, 2005)
9. I.Hafidi, On the existence of equilibrium positions in lubricated mechanisms. Ph.D. Thesis, Insa de Lyon, 2005
10. E. Marusic-Paloka, M. Starcevic, Rigorous justification of the Reynolds equations for gas lubrication, in *Comptes Rendus Mecanique*, Paris (2005), p. 333
11. E. Marusic-Paloka, M. Starcevic, Derivation of Reynolds equation for gas lubrication via asymptotic analysis of the compressible Navier-Stokes system, in *Nonlinear Analysis, Real Word Applications* (2009)
12. I. Hafidi, An investigation of the singular perturbation problems for the compressible Reynolds equation **22**(1–2), 177–200 (2009)

Heart Failure Prediction Using Supervised Machine Learning Algorithms

Soufiane Lyaqini and Mourad Nachaoui

Abstract The heart is the second most important organ of the human body after the brain, and it circulates blood and supplies all of the body's organs. Diseases associated with the heart and blood vessels involve heart failure, coronary heart disease, and rheumatic heart disease. Heart failure is the last stage of all heart diseases and the leading cause of morbidity and death among cardiac patients. With the development of machine learning, it is now feasible to predict diseases and avert life-threatening conditions. Several machine learning methods may be utilized to develop illness prediction models. These techniques and algorithms are used directly to a dataset to construct models or derive vital conclusions. This paper develops intelligent models by applying machine learning techniques to the heart failure dataset containing information about 299 patients from Faisalabad Heart Institute and Faisalabad Allied Hospital (Punjab, Pakistan). The empirical results obtained using several performance evaluation metrics demonstrate that the proposed model effectively detects coronary heart disease.

Keywords SVM · Heart failure · Smooth hinge loss · Adam algorithm

1 Introduction

Compared to the brain which is the most important organ in the human body the heart is the next most important organ it circulates blood and supplies all of the body's one. Heart disease is an illness that affects the main system of blood vessels.

S. Lyaqini
Ecole Nationale des Sciences Appliquees, LAMSAD Laboratory, Hassan First University of Settat, Berrechid, Morocco
e-mail: lyaqini.soufiane@gmail.com

M. Nachaoui (✉)
Equipe de Mathématiques et Interactions, FST Béni-Mellal, Université Sultan Moulay Slimane, Béni-Mellal, Maroc
e-mail: nachaoui@gmail.com

A. Laghrib et al. (eds.), *New Trends of Mathematical Inverse Problems and Applications*, Springer Proceedings in Mathematics & Statistics 428,
https://doi.org/10.1007/978-3-031-33069-8_4

This makes the heart and blood flow not work right. The number of people suffering from heart disease is increasing in today's world. Every year, a substantial number of individuals die from heart disease around the world [1–4]. Diseases associated to the heart and blood vessels involve: heart failure, coronary heart disease, and rheumatic heart disease. Heart failure, which is the last stage of all heart diseases and the leading cause of morbidity and death among cardiac patients. The danger and death of heart failure when it occurs in the human body must be predicted so that it can be recognized in advance and handled or therapy administered as soon as possible to reduce mortality and prolong patient survival. With the development of machine learning, it is now feasible to predict diseases and avert life-threatening conditions. Several machine learning methods may be utilized to develop illness prediction models [5–7]. The techniques and algorithms may be used directly to a dataset, in order to construct models or derive key conclusions [8–10].

In this paper, enhanced support vector machine algorithms will be used to predict heart failure patients. Then, researchers often choose a square loss function to construct a support vector machine model [9, 11, 12]. Nonetheless, when resolving classification issues with outliers, the square loss function tends to overstate the effect of outliers and makes prediction algorithms fragile [13]. To address this issue, researchers developed hinge loss, which decreases the influence of outliers on algorithm performance. This is because the hinge loss is less noise than square one [14–19]. Then, the support vector machine algorithms is rewritten as a regularized optimization problem, with the fidelity term being the max function, and the hypothisis space is a reproducing kernel Hilbert space [20]. However, since the fidelity term of the resulting minimization problem is not differentiable. To deal with this problem, we opt for smooth approximation techniques, which transform the max function into a smoothed one [21–23]. Thus, the resulting minimization problem is solved directly using Adam's algorithm [16, 24]. Lastly, we present several numerical validations of the proposed method and a comparison to existing machine learning algorithms. In particular, the obtained numerical results demonstrate that the proposed technique is an effective and valuable tool for predicting patients with heart failure.

This paper is organized as follows: Section 2 introduces the setting of the classification problem and its transformation into a minimization one, using smooth hinge loss function and the optimization approach. Section 3 presents the experimental results. Section 4 summarizes the conclusion.

2 Background and the Setting of the Problem

The hinge loss is a function that is commonly used for training classifiers such as the SVM. The Hinge loss function is defined as follows (Fig. 1):

$$L(u) = \max(u, 0).$$

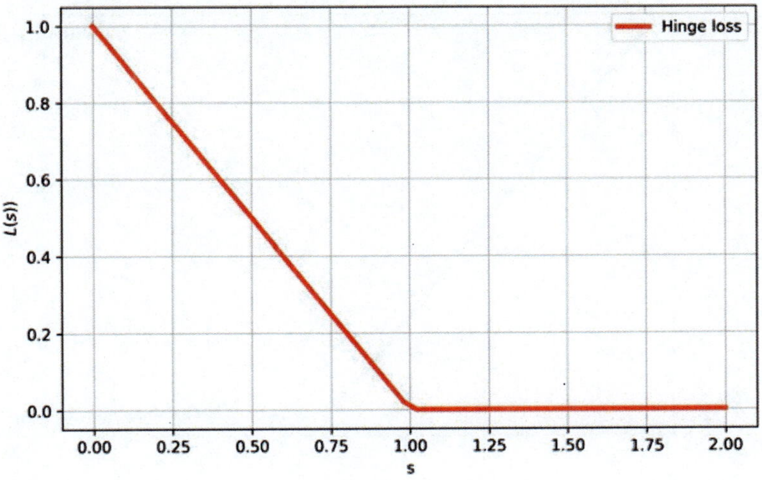

Fig. 1 Hinge loss function

The hinge loss function is not differentiable and hence cannot be employed with optimization methods that require the objective function to be differentiable. To overcome this limitation, a smooth function that is twice differentiable and convex can be used to accurately approximate the hinge loss function [15, 22] defined as follows: For all $s \in \mathbb{R}$, a smooth function is used to approximate the max function (Fig. 2).

$$\max(s, 0) = \left(s + \frac{1}{\gamma} \log\left[1 + e^{(-\epsilon s)}\right]\right), \qquad \forall \epsilon > 0.$$

We first note that the loss function is constantly a true function of only one variable s, with $s = 1 - \omega y$.

Let us denote by $L_\epsilon(s)$ the smoothed loss function with parameter ϵ of $L(s) = \max(s, 0)$ defined by

$$L_\epsilon(s) = \left(s + \frac{1}{\epsilon} \log\left[1 + e^{(-\epsilon s)}\right]\right), \qquad \forall \epsilon > 0.$$

The lemma 1 bounds the difference between hinge loss function L, and its smooth approximation L_ϵ.

Lemma 1 *For all $s \in \mathbb{R}$, then*

$$|L_\epsilon(s) - L(s)| \leq \frac{\log(2)}{\epsilon} \tag{1}$$

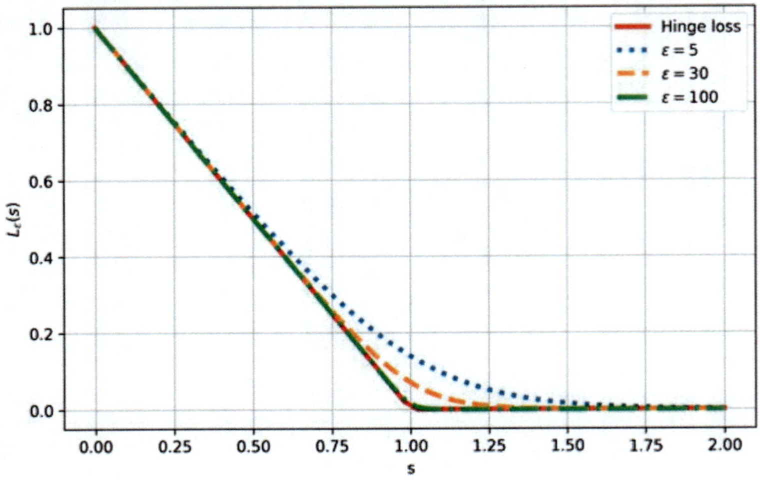

Fig. 2 Smoothed hings loss

Proof For $s > 0$, we have that $L(s) = s$, then

$$\begin{aligned}|L_\epsilon(s) - L(s)| &= |s + \frac{1}{\epsilon}\log\left[1 + e^{(-\epsilon s)}\right] - s| \\ &= \frac{1}{\epsilon}\log\left[1 + e^{(-\epsilon s)}\right] \\ &\leq \frac{\log(2)}{\epsilon}.\end{aligned}$$

Since $s \mapsto \frac{1}{\epsilon}\log\left[1 + e^{(-\epsilon s)}\right]$ is an decreasing function and $s \geq 0$.

For $s \leq 0$, we have that $L(s) = 0$, then

$$\begin{aligned}|L_\epsilon(s) - L(s)| &= |L_\epsilon(s)| \\ &\leq L_\epsilon(0) = \frac{\log(2)}{\epsilon}.\end{aligned}$$

Since L_ϵ is an increasing function and $s \leq 0$, then $L_\epsilon(s) \leq L_\epsilon(0)$. □

For the binary classification in $\mathbb{R}^n$, The training set can be expressed as follows:

$$\mathcal{T} = \left\{\left(x_j, y_j\right)\right\}_{j=1}^{n}$$

where $\boldsymbol{x}_j \in \mathbb{R}^n$ and $y_j \in \{-1, 1\}$, let $\boldsymbol{\theta} = \left(\boldsymbol{w}^T, b\right)^T$, a SVM model tries to seek a decision hyperplane $f(x) = \boldsymbol{w}^T\Phi(\boldsymbol{x}) + b = 0$, that separates two classes of data to the greatest extent. For the training set, the smooth SVM problem can be written as a convex optimization problem in (2)

$$\min_{\boldsymbol{\theta}} C\|\boldsymbol{w}\|^2 + C\sum_{j=1}^{n} L_{\epsilon}\left(1 - y_j f(x_j)\right) \tag{2}$$

where $\Phi\left(\boldsymbol{x}_j\right)$ maps $\boldsymbol{x}_j$ into a higher-dimensional space Y, $C > 0$ is the regularization parameter [25, 26], and $L_{\epsilon}(s)$ is the smoothed Hinge loss.

When the kernel method is used, we can solve the minimization problem (2) in a wider reproducing kernel Hilbert space (RKHS) $\mathcal{K}$ as

$$\min_{f \in \mathcal{K}} C\|f\|_{\mathcal{K}}^2 + \sum_{j=1}^{n} L_{\epsilon}\left(1 - y_j f(x_j)\right) \tag{3}$$

In RKHS, the Representer Theorem [27] shall be valid, then the solution of the minimization problem (3) is given as follows:

$$f^* = \sum_{j=1}^{n} \beta_j K\left(\boldsymbol{x}_j, \cdot\right),$$

where K is a kernel function in RKHS and the coefficient $\beta_j \in \mathbb{R}$.

Consequently, using the Representer Theorem in RKHS, we can further express the primal problem of the problem (3) as

$$\min_{\boldsymbol{\beta}, b} C\boldsymbol{\beta}^{\mathrm{T}}\boldsymbol{K}\boldsymbol{\beta} + \sum_{j=1}^{n} L_{\epsilon}\left(1 - (y_j, \boldsymbol{\beta}^T\boldsymbol{K}_j + b)\right) \tag{4}$$

where K is the Gram matrix derived from associated kernel function $\boldsymbol{K}_{ij} = K\left(\boldsymbol{x}_i, \boldsymbol{x}_j\right)$, $\boldsymbol{K}_j$ represents the i th row of $\boldsymbol{K}$ and $\boldsymbol{\beta}^T = (\beta_1, \beta_2, \ldots, \beta_n) \in \mathbb{R}^n$.

Many nonlinear optimization methods can be used to solve the problem (4). The gradient-type methods are very attractive to solve this kind of optimization problem. The Adam method is one of the very effective gradient-type methods due to its simplicity and low storage [24].

3 Experimental Results

In this section, we use our method to predict the presence of heart disease in a patient using a heart failure dataset from the UCI machine learning repository [28].

3.1 Dataset Description

This dataset was obtained from the UCI machine learning repository [28] and comprised medical records for 299 patients from the Faisalabad Institute of Cardiology and the Faisalabad Allied Hospital (Punjab, Pakistan). It contains data on 105 females and 194 males with Left Ventricular Systolic Dysfunction (LVSD) categorized by the New York Heart Association as phase three or stage 4 HF (NYHA). The patients' ages varied between 40 and 95 years, and the follow-up period ranged from 4 to 285 days. The dataset comprises 13 characteristics measured throughout the patients' hospital follow-up. Table 1 shows the properties within the data set and Fig. 3 show the correlation between the features.

- Hypertension or high blood pressure affects 35% of the population, while normal blood pressure affects 65%.
- Diabetes affects 42% of the population, whereas 58% are not.
- 43% of the population under study had anemic symptoms, while 57% does not.
- Males make up 65% of the population, while females make up 35%.
- 32% of the population smokes, while the remaining 68% do not.
- 32.11% of the cases died, while 67.89% of the cases survived the disease (Fig. 4).

Table 1 Description of the dataset

Attribute	Description
Age	Age of the patient
Anaemia	Decrease of red blood cells or hemoglobin
High blood pressure (Mcg/L)	If the patient has hypertension
Creatinine phosphokinase	Level of the Cpk enzyme in the blood
Diabetes	If the patient has diabetes
Ejection fraction (percentage)	Volume of blood ejected from the left ventricle in each contraction
Sex	Man or woman
Platelets (kiloplatelets/mL)	Numeric platelets count in the blood
Serum creatinine (Mg/dL)	Level of creatinine in the blood
Serum sodium (mEq/L)	Level of sodium in the blood
Smoking	If the patient has a smoking habit
Time (days)	Follow-up period
Death event (target)	If the patient died during the follow-up period

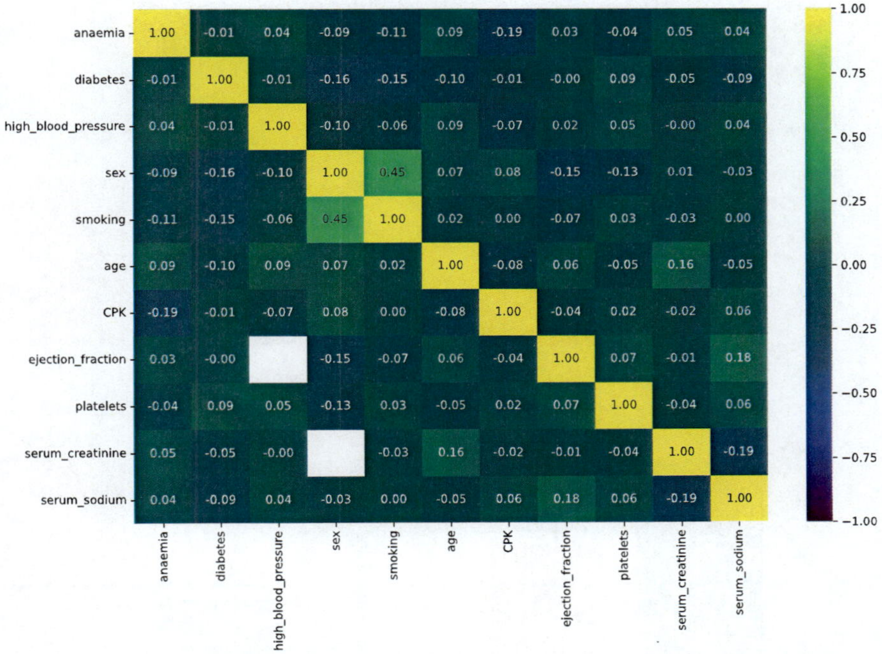

Fig. 3 Heatmap of correlation matrix

The age range of the considered population is [40–95], with substantial peaks in population density at various age intervals at [44–46], [50–52], [60–62] (highest density), [64–66], and [70–72]. The complete cast may be seen in Fig. 5.

According to the study's findings on the effect of age on survival rates: The survival rate is higher among people aged 50 to 70. Furthermore, the risk of dying from heart failure is ubiquitous throughout all age groups, with the highest risk occurring at the age of 60. After 80, the chances of survival are drastically reduced. Figure 6 illustrates the chart containing these information.

In total, 44.1% of the male population survived the condition, while 20.7% died. Similarly, 23.7% of the female population survived, whereas 11.4% died from heart failure. Figure 7 illustrates the chart containing these information.

3.2 *Results and Discussions*

Several experiments were carried out using our algorithm, SVC, Decision trees, Random Forest, XGBoost, and LGBM models [10, 29–31]. The classification algorithms' performance was evaluated in terms of sensitivity, specificity, and prediction

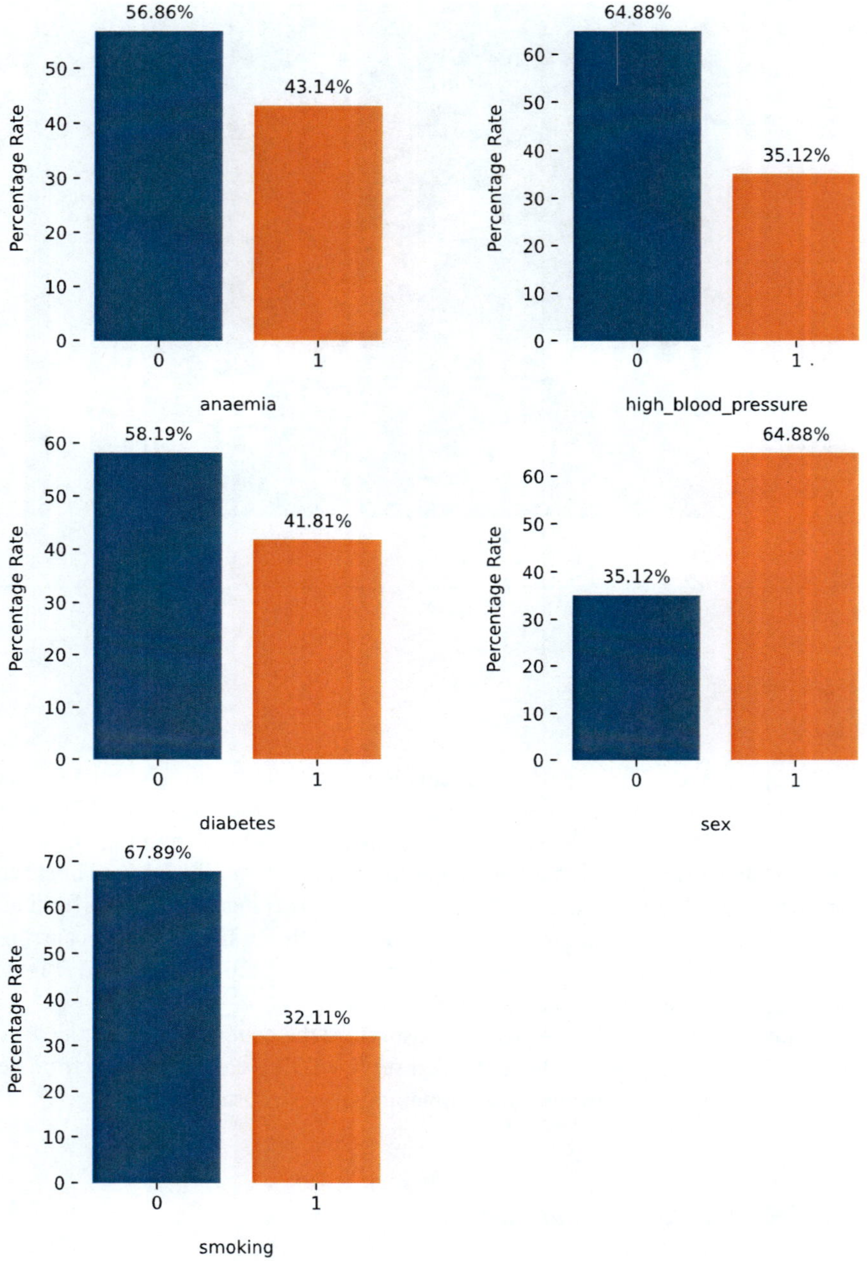

Fig. 4 Statistical information of categorical features

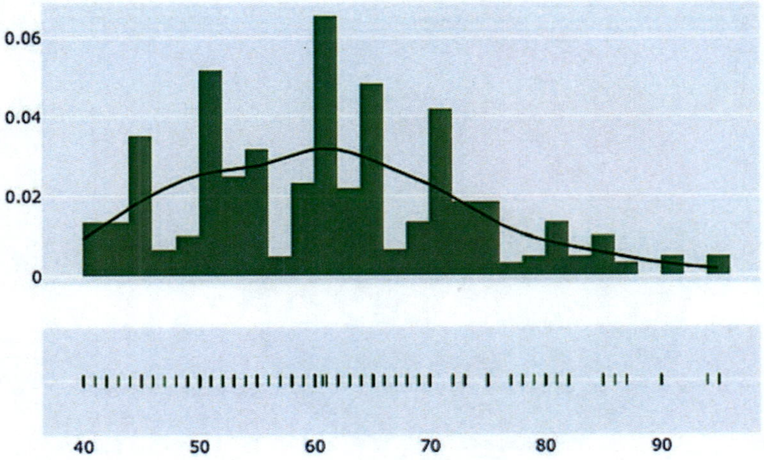

Fig. 5 Distribution of age

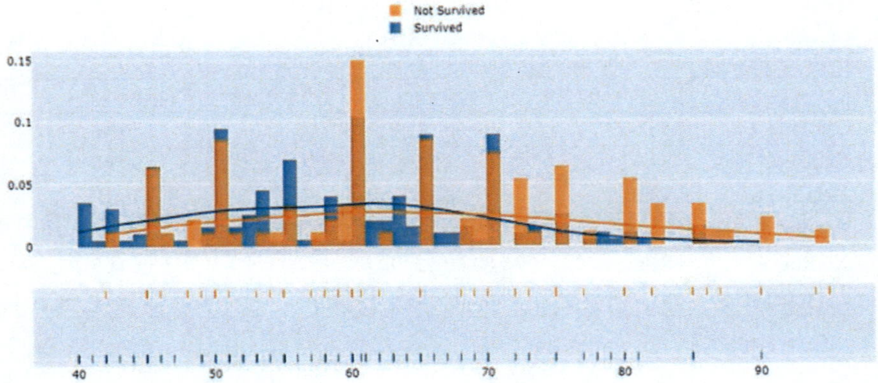

Fig. 6 Death event distribution

accuracy. Table 2 presents the results of the algorithms used to predict heart failure. It is noted that the performance of the proposed algorithm is very superior as compared to other models. Table 2 shows that our model outperforms most of the assessed performance parameters. Figure 8, depicts the proposed approach's matrix confusion and Roc Curves.

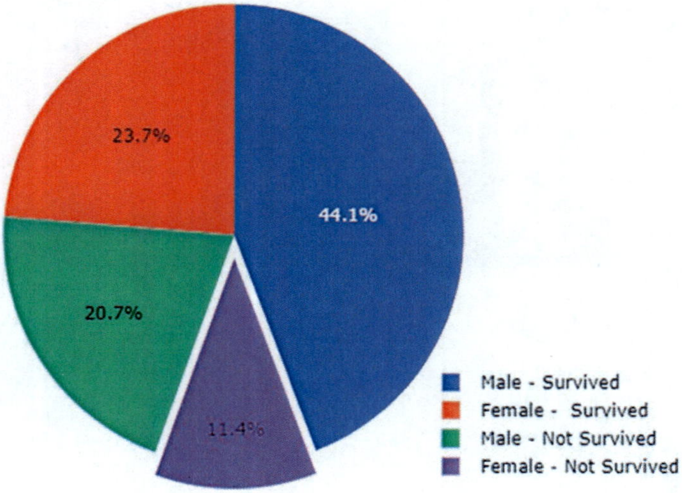

Fig. 7 Distribution of age

Table 2 Performance of the used algorithm on heart failure dataset

	$F1$ score	Accuracy	Precision	Recall
Our model	86.66	86.66	86.66	86.66
SVM	88.37	83.33	86.36	90.48
Decision tree	78.05	70.00	80.00	76.19
Random forest	85.00	80.00	89.47	80.95
XGBoost	85.00	80.00	89.47	80.95
Light GBM	85.71	80.00	85.71	85.71

3.3 Feature Importance

In this section, an extra feature ranking analysis was carried out to assess the significance of the factors based on the amount of mutual information each feature has with the DEATH EVENT.

Figure 9 depicts the average influence of the selected features on our model. Our model provides the highest serum creatinine, ejection fraction, and age ratings. Hence, Feature ratings will assist the physician in interpreting the model's outputs, contributing to developing a better AI ecosystem in the health sector.

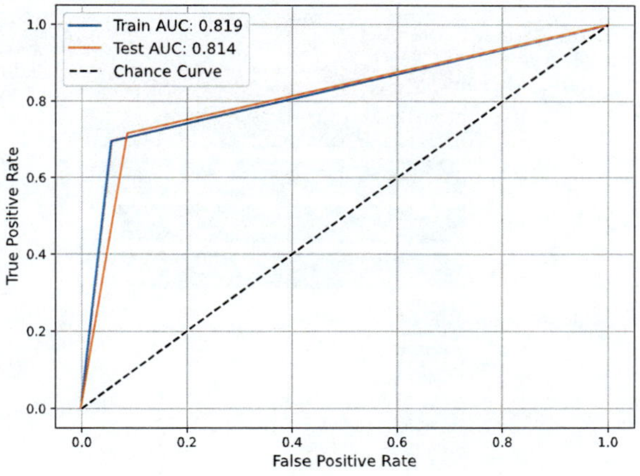

(a)

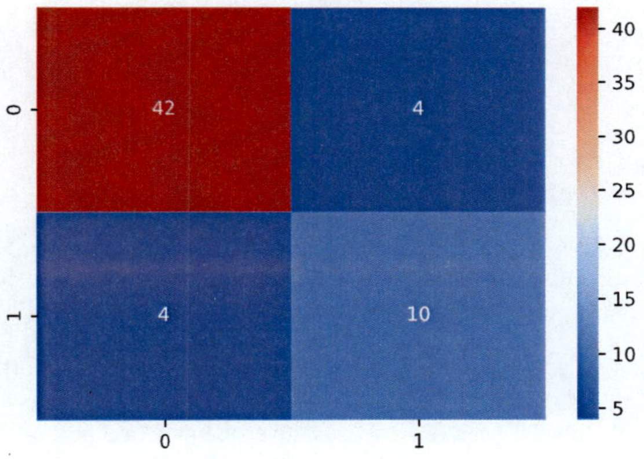

(b)

Fig. 8 **a** Roc Curves for the proposed approach. **b** Confusion matrix for the proposed approach

4 Conclusion

In this study, we propose a robust classification approach based on the smoothed hinge loss function, which is solved using Adam's algorithm. This research aimed to determine the best ML model for detecting heart failure using the heart failure dataset from Faisalabad Heart Institute and Faisalabad Allied Hospital (Punjab, Pakistan). The experiments show that our proposed approach detects heart failure more effectively than SVC, Decision trees, Random Forest, XGBoost, and LGBM models. Following that, the prediction model developed as a result of the research can be employed to create a mobile application that will allow people to control their health and thus enable early diagnosis of heart disease.

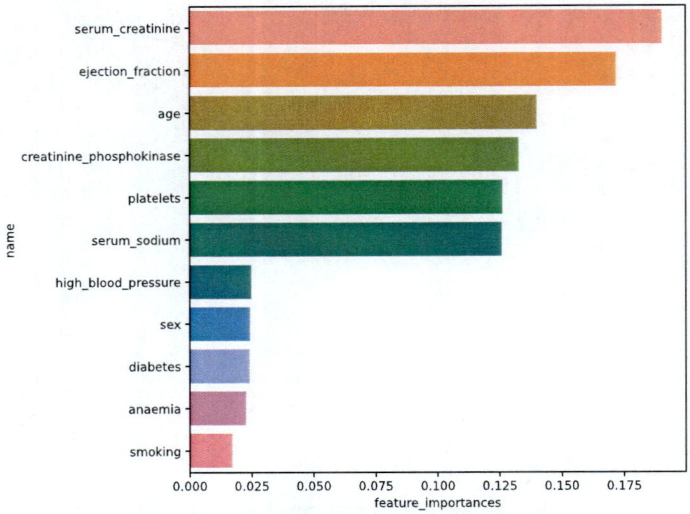

Fig. 9 Feature importance

References

1. C.W. Tsao, A.W. Aday, Z.I. Almarzooq, A. Alonso, A.Z. Beaton, M.S. Bittencourt, A.K. Boehme, A.E. Buxton, A.P. Carson, Y. Commodore-Mensah et al., Heart disease and stroke statistics-2022 update: a report from the american heart association. Circulation **145**(8), e153–e639 (2022)
2. C.M. Otto, R.A. Nishimura, R.O. Bonow, B.A. Carabello, J.P. Erwin III., F. Gentile, H. Jneid, E.V. Krieger, M. Mack, C. McLeod et al., 2020 acc/aha guideline for the management of patients with valvular heart disease: executive summary: a report of the american college of cardiology/american heart association joint committee on clinical practice guidelines. J. Am. Coll. Cardiol. **77**(4), 450–500 (2021)
3. S. Coffey, R. Roberts-Thomson, A. Brown, J. Carapetis, M. Chen, M. Enriquez-Sarano, L. Zühlke, B.D. Prendergast, Global epidemiology of valvular heart disease. Nat. Rev. Cardiol. **18**(12), 853–864 (2021)
4. A. Timmis, P. Vardas, N. Townsend, A. Torbica, H. Katus, D. De Smedt, C.P. Gale, A.P. Maggioni, S.E. Petersen, R. Huculeci et al., European society of cardiology: cardiovascular disease statistics 2021. Eur. Heart J. **43**(8), 716–799 (2022)
5. D.L.S. Jalligampala, R. Lalitha, T. Ramakrishnarao, K.R. Mylavarapu, K. Kavitha, Efficient classification of heart disease forecasting by using hyperparameter tuning, in *Applications of Artificial Intelligence and Machine Learning* (Springer, 2022), pp. 115–125
6. A. Pasteur-Rousseau, J.-F. Paul, Intelligence artificielle et téléradiologie en imagerie cardiaque en coupe, in *Annales de Cardiologie et d'Angéiologie*, vol. 70 (Elsevier, 2021), pp. 339–347
7. D. Chicco, G. Jurman, Machine learning can predict survival of patients with heart failure from serum creatinine and ejection fraction alone. BMC Med. Inform. Decis. Making **20**(1), 1–16 (2020)
8. R. Bharti, A. Khamparia, M. Shabaz, G. Dhiman, S. Pande, P. Singh, Prediction of heart disease using a combination of machine learning and deep learning, in *Computational Intelligence and Neuroscience 2021* (2021)

9. M. Kavitha, G. Gnaneswar, R. Dinesh, Y.R. Sai, R.S. Suraj, Heart disease prediction using hybrid machine learning model, in *2021 6th International Conference on Inventive Computation Technologies (ICICT)* (IEEE, 2021), pp. 1329–1333
10. S. Mishra, A comparative study for time-to-event analysis and survival prediction for heart failure condition using machine learning techniques. J. Electron. Electromed. Eng. Med. Inform. **4**(3), 115–134 (2022)
11. M.M. Ahsan, Z. Siddique, Machine learning-based heart disease diagnosis: A systematic literature review. Artif. Intell. Med. 102289 (2022)
12. M. Nachaoui, L. Afraites, A. Hadri, A. Laghrib, A non-convex non-smooth bi-level parameter learning for impulse and gaussian noise mixture removing. Commun. Pure Appl. Anal. **21**(4), 1249 (2022)
13. X. Zheng, L. Zhang, L. Yan, Ctsvm: a robust twin support vector machine with correntropy-induced loss function for binary classification problems. Inf. Sci. **559**, 22–45 (2021)
14. J. Luo, H. Qiao, B. Zhang, Learning with smooth hinge losses. Neurocomputing **463**, 379–387 (2021)
15. S. Lyaqini, M. Nachaoui, Identification of genuine from fake banknotes using an enhanced machine learning approach, in *International Conference on Numerical Analysis and Optimization Days* (Springer, 2021), pp. 59–70
16. S. Lyaqini, M. Nachaoui, A. Hadri, An efficient primal-dual method for solving non-smooth machine learning problem. Chaos Solitons Fractals **155**, 111754 (2022)
17. S. Lyaqini, M. Nachaoui, Diabetes prediction using an improved machine learning approach. Math. Model. Comput. **8**(2021), 726–735. https://doi.org/10.23939/mmc2021.04.726
18. A. Hadri, M. Nachaoui, A. Laghrib, A. Chakib, L. Afraites, A primal-dual approach for the robin inverse problem in a nonlinear elliptic equation: The case of the *l1 l2* cost functional. J. Inverse Ill-Posed Prob. (2022) [cited 2022-10-23]. https://doi.org/10.1515/jiip-2019-0098
19. M. Nachaoui, A. Laghrib, An improved bilevel optimization approach for image super-resolution based on a fractional diffusion tensor. J. Franklin Inst. **359**(13), 7165–7195 (2022)
20. B. Ghojogh, A. Ghodsi, F. Karray, M. Crowley, Reproducing kernel hilbert space, mercer's theorem, eigenfunctions, nyström method, and use of kernels in machine learning: Tutorial and survey, arXiv preprint arXiv:2106.08443 (2021)
21. S. Lyaqini, M. Quafafou, M. Nachaoui, A. Chakib, Supervised learning as an inverse problem based on non-smooth loss function. Knowl. Inf. Syst. **62**(8), 3039–3058 (2020)
22. S. Lyaqini, M. Nachaoui, M. Quafafou, Non-smooth classification model based on new smoothing technique, in *Journal of Physics: Conference Series*, vol. 1743 (IOP Publishing, 2021), p. 012025
23. M. Nachaoui, L. Afraites, A. Laghrib, A regularization by denoising super-resolution method based on genetic algorithms. Sig. Proces. Image Commun. **99**, 116505 (2021)
24. S.H. Haji, A.M. Abdulazeez, Comparison of optimization techniques based on gradient descent algorithm: A review. PalArch's J. Archaeol. Egypt/Egyptol. **18**(4), 2715–2743 (2021)
25. L. Afraites, A. Hadri, A. Laghrib, M. Nachaoui, *A non-convex denoising model for impulse and gaussian noise mixture removing using bi-level parameter identification* (Inverse Prob, Imaging, 2022)
26. L. Afraites, A. Atlas, Parameters identification in the mathematical model of immune competition cells. J. Inverse Ill-Posed Prob. **23**(4), 323–337 (2015)
27. B. Schölkopf, R. Herbrich, A.J. Smola, A generalized representer theorem, in *International Conference on Computational Learning Theory* (Springer, 2001), pp. 416–426
28. https://archive.ics.uci.edu/ml/datasets/heart+failure+clinical+records
29. K.K. Yun, S.W. Yoon, D. Won, Prediction of stock price direction using a hybrid ga-xgboost algorithm with a three-stage feature engineering process. Expert Syst. Appl. **186**, 115716 (2021)
30. P. Palimkar, R.N. Shaw, A. Ghosh, Machine learning technique to prognosis diabetes disease: random forest classifier approach, in *Advanced Computing and Intelligent Technologies* (Springer, 2022), pp. 219–244
31. R. Rivera-Lopez, J. Canul-Reich, E. Mezura-Montes, M.A. Cruz-Chávez, Induction of decision trees as classification models through metaheuristics. Swarm Evol. Comput. **69**, 101006 (2022)

Optimization Method for Estimating the Inverse Source Term in Elliptic Equation

M. Srati, A. Oulmelk, and L. Afraites

Abstract In this work, we investigate an inverse source problem for determining the unknown source term in one and two dimensional space of a linear elliptic equation. First, the inverse problem is formulated into an optimization problem with the Tikhonov regularization method. Then, the existence and uniqueness of the solution for the direct problem are proved. Second, the existence of the optimal solution is proved. Then, the convexity of the optimization problem is shown in order to ensure the uniqueness of the optimal solution. Moreover, the conjugate gradient method is applied to reconstruct the source term. Finally, to show the efficiency of the suggested approach, we give some numerical results in one and two dimensional space.

Keywords Inverse source problem · Optimal control problem · Tikhonov regularization method and conjugate gradient method

1 Introduction

An inverse problem is a mathematical framework that is used in many areas of science and engineering[1, 2, 4–14], namely biology, physics, chemistry and economics to obtain information about a physical object or system from observed measurements. However, various theory and numerical methods have been developed for solving an inverse elliptic partial differential equations (PDEs) from measured data inside or on the boundary of the domain [15–26]. For Example, the authors of [27] established uniqueness and stability of the interior inverse source problem for the Helmholtz equation from boundary Cauchy data. In [28] they applied a finite difference method for the approximate solution of the inverse problem for the multidimensional elliptic equation with overdetermination. In [29] they studied the inverse problem of finding a source term in a second-order elliptic equation defined on a finite interval. For more studies about the inverse problems of elliptic equations, we refer readers to [30–34].

M. Srati · A. Oulmelk · L. Afraites (✉)
EMI FST Beni-Mellal, Sultan Moulay Slimane University, Beni-Mellal, Morocco
e-mail: l.afraites@usms.ma

A. Laghrib et al. (eds.), *New Trends of Mathematical Inverse Problems and Applications*, Springer Proceedings in Mathematics & Statistics 428,
https://doi.org/10.1007/978-3-031-33069-8_5

See also the work [3, 35–37] for the inverse source problem governed by a fractional time model.

In this work, we examine an inverse source problem for an elliptic partial differential equation define by

$$\begin{cases} -\Delta u(x) + a(x)u(x) = f(x), \ x \in \mathbf{\Omega}, \\ \frac{\partial u}{\partial n}(x) + b(x)u(x) = g(x), \qquad x \in \partial\mathbf{\Omega}, \end{cases} \tag{1}$$

where $\mathbf{\Omega}$ is bounded domain in $\mathbb{R}^n$ $(n \geq 1)$, with sufficient smooth boundary $\partial\mathbf{\Omega}$, $f \in L^2(\mathbf{\Omega})$, $g \in L^2(\partial\mathbf{\Omega})$, the coefficients $a \in L^\infty(\mathbf{\Omega})$, $b \in L^\infty(\partial\mathbf{\Omega})$ and are assumed to satisfy the following conditions

$$\underline{a} \leq a(x) \leq \overline{a} \ \text{ in } \mathbf{\Omega}, \text{ and } \underline{b} \leq b(x) \leq \overline{b} \ \text{ in } \partial\mathbf{\Omega}, \tag{2}$$

here $\underline{a}$, $\overline{a}$, $\underline{b}$, $\overline{b}$ are known positive constants, $\frac{\partial u}{\partial n}$ denotes the normal derivative of u on $\partial\mathbf{\Omega}$.

If all the parameters f, g, a, b are given appropriately, then the problem (1) is called the state problem or direct problem. While if one of these parameters is unknown, we called it in inverse problem.

The inverse problem we aim to treat in this work is the identification of the unknowns source term f in the Eq. (1) from the following measurements

$$u(x) = \phi(x), \quad .a.e \ \ x \in \mathbf{\Omega}, \tag{3}$$

where $\phi \in L^2(\mathbf{\Omega})$ is given.

One of the most frequently used approaches to solve the inverse problem is the optimization approach. For this, we reformulate the inverse problem (1)–(3) in the following PDE-constrained optimization one with Tikhonov regularization. The optimization model is given as follow

$$\mathbb{J}(\widehat{f}) = \min_{f \in \mathcal{A}} \mathbb{J}(f), \tag{4}$$

where

$$\mathbb{J}(f) = \frac{1}{2}\int_{\mathbf{\Omega}} |u(x) - \phi(x)|^2 dx + \frac{\lambda}{2}\int_{\mathbf{\Omega}} |f(x)|^2 dx, \tag{5}$$

with u is the solution of the direct problem (1), and the admissible set $\mathcal{A}$ is defined by

$$\mathcal{A} = \left\{ f \in L^2(\mathbf{\Omega}) \ ; \ \underline{f} \leq f(x) \leq \overline{f} \ \text{ for almost every } x \in \mathbf{\Omega} \right\}.$$

with $\underline{f}$, $\overline{f}$ are known constants, and $\lambda > 0$ is a regularization parameter (see [5, 13, 38, 39, 42–44] for the choice of λ).

This work is organized as follows. In Sect. 2, we give the variational formulation of the direct problem. Then, the existence and the uniqueness results of the direct problem are shown. The existence of the solution for the optimization problem is given in Sect. 3. In Sect. 4, we prove the convexity of the control problem to ensure the uniqueness of the minimizer. In Sect. 5, we present the suggested conjugate gradient algorithm. The numerical results are presented in Sect. 6 in the dimension one and two.

2 Variational Formulation and a Priori Estimates

In this section we introduce the variational formulation of the problem (1) and a priori estimates of the weak solution u. Then, the variational formulation corresponding to the problem (1) is expressed in the following manner

$$\begin{cases} \text{Find } u \in H^1(\mathbf{\Omega}) \text{such that,} \\ \mathbb{B}(u, v) = \mathbb{L}(v), \quad \forall\, v \in H^1(\mathbf{\Omega}), \end{cases} \tag{6}$$

where $\mathbb{B}$ is a bilinear form defined by

$$\mathbb{B}(u, v) = \int_{\mathbf{\Omega}} \nabla u(x) \nabla v(x) dx + \int_{\mathbf{\Omega}} a(x) u(x) v(x) dx + \int_{\partial\mathbf{\Omega}} b(x) u(x) v(x) dx,$$

and $\mathbb{L}$ is a linear form defined by

$$\mathbb{L}(v) = \int_{\mathbf{\Omega}} f(x) v(x) dx + \int_{\partial\mathbf{\Omega}} g(x) v(x) dx.$$

The following theorem show the existence and uniqueness of the solution for the problem (6).

Theorem 1 *Assume that $a \in L^\infty(\mathbf{\Omega})$, $b \in L^\infty(\partial\mathbf{\Omega})$ are two strictly positive functions, $f \in \mathcal{A}$ and $g \in L^2(\partial\mathbf{\Omega})$. Then, there existe a unique weak solution $u \in H^1(\mathbf{\Omega})$ of the problem (6), and we have the following a priori estimations*

$$\|u\|_{H^1(\mathbf{\Omega})} \leq C\big(\|f\|_{L^2(\mathbf{\Omega})} + \|g\|_{L^2(\partial\mathbf{\Omega})}\big), \tag{7}$$

where C is a positive constant.

Proof To prove this theorem, we will use the Lax-Milgram theorem. For this, we show that the bilinear form $\mathbb{B}$ is continuous and elliptic on $H^1(\mathbf{\Omega})$, and the linear form $\mathbb{L}$ is continuous on $H^1(\mathbf{\Omega})$.

- Continuity of the bilinear form $\mathbb{B}$:
  For any $u, v \in H^1(\mathbf{\Omega})$ and by using the Cauchy-Schwarz inequality, we have

$$|\mathbb{B}(u,v)| \leq \|\nabla u\|_{L^2(\mathbf{\Omega})}\|\nabla v\|_{L^2(\mathbf{\Omega})} + \overline{a}\|u\|_{L^2(\mathbf{\Omega})}\|v\|_{L^2(\mathbf{\Omega})} + \overline{b}\|u\|_{L^2(\partial\mathbf{\Omega})}\|v\|_{L^2(\partial\mathbf{\Omega})} \tag{8}$$

Since $u, v \in\in H^1(\mathbf{\Omega})$, then by the trace theorem we have

$$|\mathbb{B}(u,v)| \leq \|u\|_{H^1(\mathbf{\Omega})}\|v\|_{H^1(\mathbf{\Omega})} + \overline{a}\|u\|_{H^1(\mathbf{\Omega})}\|v\|_{H^1(\mathbf{\Omega})} + C_1C_2\overline{b}\|u\|_{H^1(\mathbf{\Omega})}\|v\|_{H^1(\mathbf{\Omega})} \tag{9}$$

Hence, we obtain

$$|\mathbb{B}(u,v)| \leq C\|u\|_{H^1(\mathbf{\Omega})}\|v\|_{H^1(\mathbf{\Omega})} \tag{10}$$

where $C = 1 + \overline{a} + C_1C_2\overline{b}$, which gives that $\mathbb{B}$ is continuous.

- Ellipticity of the bilinear form $\mathbb{B}$:
  For any $u \in H^1(\mathbf{\Omega})$, we have

$$\begin{aligned}\mathbb{B}(u,u) &\geq \|\nabla u\|^2_{L^2(\mathbf{\Omega})} + \underline{a}\|u\|^2_{L^2(\mathbf{\Omega})} + \underline{b}\|u\|^2_{L^2(\partial\mathbf{\Omega})}\\ &\geq \min(1,\underline{a})\|u\|^2_{H^1(\mathbf{\Omega})} + \underline{b}\|u\|^2_{L^2(\partial\mathbf{\Omega})}\\ &\geq \min(1,\underline{a})\|u\|^2_{H^1(\mathbf{\Omega})},\end{aligned}$$

  which gives that $\mathbb{B}$ is elliptic.
- Continuity of the linear form $\mathbb{L}$:
  For any $v \in H^1(\mathbf{\Omega})$ and by using the Cauchy-Schwarz inequality, we have

$$|\mathbb{L}(v)| \leq \|f\|_{L^2(\mathbf{\Omega})}\|v\|_{L^2(\mathbf{\Omega})} + \|g\|_{L^2(\partial\mathbf{\Omega})}\|v\|_{L^2(\partial\mathbf{\Omega})} \tag{11}$$

  Sine $v \in H^1(\mathbf{\Omega})$, then by the trace theorem we have

$$\begin{aligned}|\mathbb{L}(v)| &\leq \|f\|_{L^2(\mathbf{\Omega})}\|v\|_{L^2(\mathbf{\Omega})} + C_3\|g\|_{L^2(\partial\mathbf{\Omega})}\|v\|_{H^1(\mathbf{\Omega})}\\ &\leq C_4\Big(\|f\|_{L^2(\mathbf{\Omega})} + \|g\|_{L^2(\partial\mathbf{\Omega})}\Big)\|v\|_{H^1(\mathbf{\Omega})}\end{aligned}$$

  where $C_4 = \max(1, C_3)$, which gives that $\mathbb{L}$ is continuous.

By take $u = v$ in the variational formulation (6), we have

$$\mathbb{B}(u,u) = \mathbb{L}(u). \tag{12}$$

From the ellipticity of $\mathbb{B}$ and the continuity of $\mathbb{L}$, we obtain the following estimate

$$\|u\|_{H^1(\mathbf{\Omega})} \leq C\Big(\|f\|_{L^2(\mathbf{\Omega})} + \|g\|_{L^2(\partial\mathbf{\Omega})}\Big), \tag{13}$$

where $C = \frac{C_4}{\min(1,\underline{a})}$.
This completes the proof of the Theorem 1. □

3 Existence of the Optimization Problem

in this section, we prove that the existence result of the solution for the optimal control problem (5).

Theorem 2 *The optimization problem (4) admits at least a solution in* $\mathcal{A}$

Proof Let's $(f^{(n)})_n$ be a minimizing sequence of $\mathcal{J}$ in $\mathcal{A}$ such that

$$\lim_{n\longmapsto\infty} \mathbb{J}(f^{(n)}) = \inf_{f\in\mathcal{A}} \mathbb{J}(f). \tag{14}$$

By definition of $\mathcal{A}$, we have $(f^{(n)})_n$ is bounded in $L^2(\boldsymbol{\Omega})$ which means that there exists a subsequence, still denoted by $(f^{(n)})_n$ and $\bar{f} \in \mathcal{A}$ such that

$$f^{(n)} \rightharpoonup \bar{f} \text{weakly in} \quad L^2(\boldsymbol{\Omega}). \tag{15}$$

Let $u^{(n)} = u(f^{(n)})$ is the solution of (6) associated to $f^{(n)}$ for all $n \in \mathbb{N}$. Then we have

$$\mathbb{B}(u^{(n)}, v) = \mathbb{L}(v), \quad \forall v \in H^1(\boldsymbol{\Omega}). \tag{16}$$

By taking $v = u^{(n)}$ in (16) and using the same technique as in the Theorem 1, we obtain

$$\|u^{(n)}\|_{H^1(\boldsymbol{\Omega})} \leq C\big(\|f\|_{L^2(\boldsymbol{\Omega})} + \|g\|_{L^2(\partial\boldsymbol{\Omega})}\big), \quad \forall n \in \mathbb{N}, \tag{17}$$

which gives that $\|u^{(n)}\|_{H^1(\boldsymbol{\Omega})}$ is uniformly bounded independent of n. Then the compactness results imply that there exists a subsequence of $u^{(n)}$, also denoted by $u^{(n)}$, and $\bar{u} \in H^1(\boldsymbol{\Omega})$ such that

$$u^{(n)} \rightharpoonup \bar{u} \text{ weakly in} \quad H^1(\boldsymbol{\Omega}). \tag{18}$$

The Sobolev trace theorem and the Sobolev theorem ensures the compact embedding from $H^1(\boldsymbol{\Omega})$ into $L^2(\partial\boldsymbol{\Omega})$ and the compact embedding of $H^1(\boldsymbol{\Omega})$ into $L^2(\boldsymbol{\Omega})$. Then, there exists a subsequence, still denoted by $u^{(n)}$ such that

$$u^{(n)} \rightarrow \bar{u} \quad \text{strongly in} \quad L^2(\partial\boldsymbol{\Omega}), \tag{19}$$

and

$$u^{(n)} \rightarrow \bar{u} \quad \text{strongly in} \quad L^2(\boldsymbol{\Omega}). \tag{20}$$

Now, we prove that $\bar{u}$ is solution of (6) associated to $\bar{f}$. Let $v \in H^1(\boldsymbol{\Omega})$.
From (19), (18), and the fact that b is is bounded in $L^\infty(\partial\boldsymbol{\Omega})$, we have

$$\lim_{n\longrightarrow\infty} \int_{\partial\boldsymbol{\Omega}} bu^{(n)}v = \int_{\partial\boldsymbol{\Omega}} b\bar{u}v, \quad \lim_{n\longrightarrow\infty} \int_{\boldsymbol{\Omega}} \nabla u^{(n)}\nabla v = \int_{\boldsymbol{\Omega}} \nabla\bar{u}\nabla v. \tag{21}$$

From (20), (15), and the fact that a is is bounded in $L^{\infty}(\mathbf{\Omega})$, we have

$$\lim_{n\longrightarrow\infty}\int_{\mathbf{\Omega}} au^{(n)}v = \int_{\mathbf{\Omega}} a\bar{u}v, \quad \lim_{n\longrightarrow\infty}\int_{\mathbf{\Omega}} f^{(n)}v = \int_{\mathbf{\Omega}} \bar{f}v. \tag{22}$$

Letting n tend to ∞ in Eq. (16), we conclude that the $\bar{u}$ satisfies

$$\mathbb{B}(\bar{u}, v) = \mathbb{L}(v), \quad \forall v \in H^1(\mathbf{\Omega}), \tag{23}$$

which gives $\bar{u} = u(\bar{f})$.
Now the strong convergence of $u^{(n)}$ in $L^2(\mathbf{\Omega})$ and the lower semicontinuity of a norm imply that

$$\begin{aligned}
\mathbb{J}(\bar{f}) &= \frac{1}{2}\|\bar{u} - \phi\|^2_{L^2(\mathbf{\Omega})} + \frac{\mu}{2}\|\bar{f}\|^2_{L^2(\mathbf{\Omega})} \\
&\leq \frac{1}{2}\lim_{n\longmapsto\infty}\|u^{(n)} - \phi\|^2_{L^2(\mathbf{\Omega})} + \frac{\lambda}{2}\liminf_{n\longmapsto\infty}\|f^{(n)}\|^2_{L^2(\mathbf{\Omega})} \\
&\leq \liminf_{n\longmapsto\infty}\left(\frac{1}{2}\|u^{(n)} - \phi\|^2_{L^2(\mathbf{\Omega})} + \frac{\lambda}{2}\|f^{(n)}\|^2_{L^2(\mathbf{\Omega})}\right) \\
&\leq \liminf_{n\longmapsto\infty}\mathbb{J}(f^{(n)}) = \inf_{f\in\mathcal{A}}\mathbb{J}(f).
\end{aligned}$$

Therefore $\bar{f}$ is a minimizer of the functional $\mathbb{J}$.
This completes the proof of the Theorem 2. □

4 Convexity of the Optimization Problem

In this section, we show the convexity of the optimization problem (5) in order to prove the uniqueness of the solution for the optimization problem (5) . First, we show that the operator source to solution is Lipschitz continuity and its differentiability with respect to the source f in order to establish the convexity of the cost functional $\mathbb{J}$. For this, we define source-to-solution operator $\mathcal{K} : f \in \mathcal{A} \longrightarrow H^1(\mathbf{\Omega})$ defined by $\mathcal{K}(f) = u = u_f$ with u is the solution of the problem (6).

The following Theorem shows that the $\mathcal{K}$ is lipschitz continuous.

Theorem 3 *Let u_{f_1}, u_{f_2} the solution of the problem (6) corresponding to $f_1, f_2 \in \mathcal{A}$,respectively. Then there existe a positive constant C such that*

$$\|\mathcal{K}(f_1) - \mathcal{K}(f_2)\|_{H^1(\mathbf{\Omega})} \leq C\|f_1 - f_2\|_{L^2(\mathbf{\Omega})}. \tag{24}$$

Proof By positing $w = \mathcal{K}(f_1) - \mathcal{K}(f_2) = u_{f_1} - u_{f_2}$, then w satisfies the following problem

$$\begin{cases} -\Delta w(x) + a(x)w(x) = f_1(x) - f_2(x), \; x \in \mathbf{\Omega}, \\ \frac{\partial w}{\partial n}(x) + b(x)w(x) = 0, \qquad x \in \partial\mathbf{\Omega}. \end{cases} \tag{25}$$

The variational formulation associate to (25) is given by
find $w \in H^1(\mathbf{\Omega})$ such that

$$\mathbb{B}(w, v) = \int_{\mathbf{\Omega}} \big(f_1(x) - f_2(x)\big)v(x)dx, \quad \forall v \in H^1(\mathbf{\Omega}). \tag{26}$$

By taking $v = w$ in (26), and using the ellipticity of $\mathbb{B}$, and the Cauchy-Schwarz inequality, we have

$$\|w\|_{H^1(\mathbf{\Omega})} \leq C\|f_1 - f_2\|_{L^2(\mathbf{\Omega})}, \tag{27}$$

where $\frac{1}{C} = \min(1, \underline{a})$.
This completes the proof of Theorem 3. □

The following Theorem shows the first Fréchet differentiability of the operator source-to-solution $\mathcal{K}$.

Theorem 4 *For any $f \in \mathcal{A}$ the operator $\mathcal{K}$ is differentiable with respect to f, if $u' = D\mathcal{K}(f)h$ then u' is unique solution to the following sensitivity problem*

$$\begin{cases} -\Delta u'(x) + a(x)u'(x) = h(x), \; x \in \mathbf{\Omega}, \\ \frac{\partial u'}{\partial n}(x) + b(x)u'(x) = 0, \qquad x \in \partial\mathbf{\Omega}, \end{cases}. \tag{28}$$

The variational formulation associate to (28) is given by
find $u' \in H^1(\mathbf{\Omega})$ such that

$$\mathbb{B}(u', v) = \int_{\mathbf{\Omega}} h(x)v(x)dx, \quad \forall v \in H^1(\mathbf{\Omega}). \tag{29}$$

Moreover, we have

$$\|D\mathcal{K}(f)\|_{H^1(\mathbf{\Omega})} \leq C, \tag{30}$$

where $C = \frac{1}{\min(1,\underline{a})}$.

Proof Let $f \in \mathcal{A}$ and $h \in L^2(\mathbf{\Omega})$ sufficiently small such as $f + h \in \mathcal{A}$. We pose $w = \mathcal{K}(f + h) - \mathcal{K}(f)$, then w satisfies the following problem

$$\begin{cases} -\Delta w(x) + a(x)w(x) = h(x), \; x \in \mathbf{\Omega}, \\ \frac{\partial w}{\partial n}(x) + b(x)w(x) = 0, \qquad x \in \partial\mathbf{\Omega}. \end{cases} \tag{31}$$

The variational formulation associate to (31) is given by
find $w \in H^1(\mathbf{\Omega})$ such that

$$\mathbb{B}(w, v) = \int_{\mathbf{\Omega}} h(x)v(x)dx, \quad \forall v \in H^1(\mathbf{\Omega}), \tag{32}$$

by compining (29) and (32), we have

$$\mathbb{B}(w - u', v) = 0, \quad \forall v \in H^1(\mathbf{\Omega}). \tag{33}$$

By taking $v = w - u'$ in (33) and by using the ellipticity of $\mathcal{B}$, we obtain

$$\|w - u'\|^2_{H^1(\mathbf{\Omega})} = 0,$$

which gives

$$\|w - u'\|_{H^1(\mathbf{\Omega})} = o(\|h\|^2_{L^2(\mathbf{\Omega})}.$$

Then, $\mathcal{K}$ is differentiable at f and $D\mathcal{K}(f)h = u'$.
Also, let $v = u'$ in (29) and using the ellipticity of $\mathcal{B}$, we get

$$\|u'\|_{H^1(\mathbf{\Omega})} \leq \frac{1}{\min(1, \underline{a})}\|h\|_{L^2(\mathbf{\Omega})}.$$

Finaly, we have

$$\|D\mathcal{K}(f)\|_{H^1(\mathbf{\Omega})} \leq \frac{1}{\min(1, \underline{a})}.$$

The following theorem gives the expression of the first derivative of the cost function $\mathbb{J}$.

Theorem 5 *Let $f \in \mathcal{A}$. then the first derivative of the cost function $\mathbb{J}$ at f is given by*

$$\mathbb{J}'(f) = p + \lambda f, \tag{34}$$

where p is the solution of the following adjoint problem

$$\begin{cases} -\Delta p(x) + a(x)p(x) = u(x) - \phi(x), \ x \in \mathbf{\Omega}, \\ \frac{\partial p}{\partial n}(x) + b(x)p(x) = 0, \qquad x \in \partial\mathbf{\Omega}. \end{cases} \tag{35}$$

Proof Let $f \in \mathcal{A}$ and $h \in L^2(\mathbf{\Omega})$ sufficiently small such as $f + h \in \mathcal{A}$. Then the derivative of the cost functional $\mathbb{J}$ with respect to f in the direction h is given by the following expression

$$\mathbb{J}'(f)h = \int_{\mathbf{\Omega}} \big(u(x) - \phi(x)\big)u'(x)dx + \lambda \int_{\mathbf{\Omega}} f(x)h(x)dx, \tag{36}$$

where u' s the solution of the sensitivity problem (28). Multiplying (35) by u' and integrating over Ω and using the Green formula and the boundary conditions, we obtain

$$-\int_{\Omega}\Delta u'(x)p(x)dx+\int_{\Omega}a(x)u'(x)p(x)dx=\int_{\Omega}\big(u(x)-\phi(x)\big)u'(x)dx.$$

Since u' is the solution of the problem (28), then we have

$$\int_{\Omega}h(x)p(x)dx=\int_{\Omega}\big(u(x)-\phi(x)\big)u'(x)dx \tag{37}$$

substitute (37) into (36), we get

$$\begin{aligned}\mathbb{J}'(f)h&=\int_{\Omega}h(x)p(x)dx+\lambda\int_{\Omega}f(x)h(x)dx\\&=\int_{\Omega}\Big(p(x)+\lambda f(x)\Big)h(x)dx,\end{aligned}$$

which gives that

$$\mathbb{J}'(f)=p+\lambda f.$$

This completes the proof of the Theorem 5. □

The convexity result of the cost functional $\mathbb{J}$ is established in the following theorem.

Theorem 6 *For any $\lambda>0$, the cost functional $\mathbb{J}$ defined in (5) is strictly convex.*

Proof Let $f,h\in\mathcal{A}$, from the above, it is clear that $\mathbb{J}$ is differentiable with respect to f, so to prove that $\mathbb{J}$ is strictly convex it is sufficient to prove the following inequality

$$\mathbb{J}(f+h)-\mathbb{J}(f)-\mathbb{J}'(f).h>0,\forall h\in\mathcal{A}. \tag{38}$$

we have

$$\begin{aligned}&\mathbb{J}(f+h)-\mathbb{J}(f)-\mathbb{J}'(f).h\\&\quad=\frac{1}{2}\|u_{f+h}-\phi\|^2_{L^2(\Omega)}+\frac{\lambda}{2}\|f+h\|^2_{L^2(\Omega)}-\frac{1}{2}\|u_f-\phi\|^2_{L^2(\Omega)}-\frac{\lambda}{2}\|f\|^2_{L^2(\Omega)}\\&\qquad-\int_{\Omega}(u_f-\phi)u'dx-\lambda\int_{\Omega}fhdx\\&\quad=\frac{1}{2}\int_{\Omega}\big(u_{f+h}+u_f-2\phi\big)\big(u_{f+h}-u_f\big)dx+\frac{\lambda}{2}\|h\|^2_{L^2(\Omega)}\\&\qquad-\int_{\Omega}(u_f-\phi)u'dx\\&\quad=\frac{1}{2}\int_{\Omega}\big(u_{f+h}-u_f\big)\big(u_{f+h}-u_f\big)dx+\int_{\Omega}\big(u_{f+h}-u_f\big)\big(u_f-\phi\big)dx\\&\qquad-\int_{\Omega}(u_f-\phi)u'dx+\frac{\lambda}{2}\|h\|^2_{L^2(\Omega)}\end{aligned}$$

$$= \underbrace{\frac{1}{2}\|u_{f+h} - u_f\|^2_{L^2(\Omega)}}_{I_1} + \underbrace{\int_\Omega \big(u_{f+h} - u_f - u'\big)\big(u_f - \phi\big)dx}_{I_2} + \frac{\lambda}{2}\|h\|^2_{L^2(\Omega)}.$$

From above, and the fact that the operator source to solution is affine, then we have $u_{f+h} - u_f = u', \ \forall h \in \mathcal{A}$, which gives

$$I_2 = 0, \quad I_1 \geq 0. \tag{39}$$

Therefore, we have

$$\mathbb{J}(f+h) - \mathbb{J}(f) - \mathbb{J}'(f).h \geq \frac{\lambda}{2}\|h\|^2_{L^2(\Omega)}.$$

Finaly, we obtain $\mathbb{J}$ is strictly convex, $\quad \forall \lambda > 0$. □

5 The Proposed Algorithm

In this section, we give the proposed algorithm based on the conjugate gradient method to search the minimizer of the cost functional $\mathbb{J}$. For this, let $f^{(k)}$ be the kth approximate solution to f. We use the following iterative scheme

$$f^{(k+1)} = f^{(k)} + \theta^{(k)} d^{(k)}, \quad \forall k \in \mathbb{N}, \tag{40}$$

where $\theta^{(k)}$ s a step size and $d^{(k)}$ is a descent direction in the kth iteration of the form

$$d^{(k)} = -\mathbb{J}'(f^{(k)}) + \zeta^{(k)} d^{(k-1)}, \tag{41}$$

where $\zeta^{(k)}$ is given by

$$\zeta^{(k)} = \frac{\displaystyle\int_\Omega \big(\mathbb{J}'(f^{(k)})\big)^2}{\displaystyle\int_\Omega \big(\mathbb{J}'(f^{(k-1)})\big)^2}, \quad \zeta^{(0)} = 0, \tag{42}$$

we have

$$\mathbb{J}(f^{(k)} + \theta^{(k)} d^{(k)}) = \frac{1}{2}\int_\Omega (u_{f^{(k)}}(x) + \theta^{(k)} w^{(k)}(x) - \phi(x))^2 dx + \frac{\lambda}{2}\int_\Omega (f^{(k)} + \theta^{(k)} w^{(k)})^2 dx, \tag{43}$$

where $w^{(k)}$ is a solution of the sensitive problem (28) with $h = d^{(k)}$

$$\frac{d\mathbb{J}}{d\theta^{(k)}} = \int_{\Omega}(u_{f^{(k)}} - \phi(x) + \theta^{(k)}w^{(k)})w^{(k)}dx + \lambda\int_{\Omega}(f^{(k)} + \theta^{(k)}w^{(k)})d^{(k)}dx = 0, \tag{44}$$

then

$$\theta^{(k)} = -\frac{\displaystyle\int_{\Omega}(u_{f^{(k)}} - \phi(x))w^{(k)}dx + \int_{\Omega} f^{(k)}d^{(k)}dx}{\displaystyle\int_{\Omega} w^{(k)}dx + \lambda\int_{\Omega} d^{(k)}dx}. \tag{45}$$

The Conjugate gradient algorithm is presented in Algorithm 1.

Algorithm 1 The Conjugate gradient algorithm for solving he minimization problem

Begin
Input: Choose an initial source $f^{(0)}, d^{(0)} = \mathbb{J}'(f^{(0)})$, the regularization parameter λ, tol, max_{itr},
Initialize Set $k = 0$.
  while $k < max_{itr}$ **do**
    Solve the direct problem (1) with $f = f^{(k)}$
    Solve the adjoint problem (35) and determine the gradient $\mathbb{J}'(f^{(k)})$.
    Calculate the conjugate coefficient $\zeta^{(k)}$ by (42) and the descent direct $d^{(k)}$ by (41).
    Solve the sensitive problem (28) with $h = d^{(k)}$ to obtain $w^{(k)}$
    Calculate the step size $\theta^{(k)}$ by (45).
    update the source term $f^{(k+1)}$ by (40).
    Check convergence : **if** $\frac{\|f^{(k+1)} - f^{(k)}\|_{L^2(\Omega)}}{\|f^{(k)}\|_{L^2(\Omega)}} \leq tol$ stop.
    Set $k = k + 1$.
  end while
return $f^{(k)}$.
End

6 Numerical Results

In this section, we present eight examples in the one-dimensional case and four examples in the two-dimensional case to test the efficiency of the conjugated gradient algorithm. We present some numerical results in order to confirm our previous theoretical results. We construct the synthetic data ϕ by fixing the unknown source term f, choosing the coefficients a,b and g, then we solve the direct problem (1) and we extract the measurement ϕ. the noisy is generated by adding a random perturbation in f by the following formula

$$f^{\sigma} = f + \sigma f.(2rand(f) - 1). \tag{46}$$

To show the accuracy of the numerical solution of the conjugate gradient method, we compute the relative error denoted by

$$e^{(k)} = \frac{\|f - f^{(k)}\|_{L^2(\mathbf{\Omega})}}{\|f\|_{L^2(\mathbf{\Omega})}}. \tag{47}$$

where $f^{(k)}$ is the source term reconstructed at the k-th iterations and f is the exact solution term. We use a finite difference method to solve the direct problem, the sensitive problem and the adjoint problem.

6.1 Results for the One-Dimensional Case

In the subsection, we provide eight cases in a one-dimensional case. The four initial functions are smooth while the final functions are nonsmooth. In numerical computation, we take $n = 1$, $\mathbf{\Omega} = (0, 2)$ and $\lambda = 10^{-6}$.

Example 1 In the first example, we consider a smooth source function as follows

$$f(x) = e^{-2x}\sin(4\pi x).$$

Example 2 In the second example, we consider a smooth source function as follows

$$f(x) = x^2(x-2)^2.$$

Example 3 In the third example, we consider a smooth source function as follows

$$f(x) = 10\sin(\pi(x-1)) + x(4 - \frac{1}{2}x^3).$$

Example 4 In the fourth example, we consider a smooth source function as follows

$$f(x) = e^{-3\sin(5\pi x)} + x(4 - \frac{1}{2}x^3).$$

Example 5 In the five example, we consider a non-smooth source function as follows

$$f(x) = \begin{cases} -2x & \text{if } 0 \le x \le 0.4, \\ -0.8 + 2x & \text{if } 0.4 \le x \le 0.8, \\ 0 & \text{if } 0.8 \le x \le 1.2, \\ 2 & \text{if } 1.2 \le x \le 1.6, \\ 2\sin(10\pi(x-0.8)(x-2)) & \text{if } 1.6 \le x \le 2.0, \end{cases}.$$

Example 6 In the six example, we consider a non-smooth source function as follows

$$f(x) = \begin{cases} 0 & \text{if} 0 \leq x \leq \frac{1}{4}, \\ -\frac{20}{3}x + \frac{16}{3} & \text{if} \frac{1}{4} \leq x \leq \frac{7}{20}, \\ \frac{20}{3}x - \frac{10}{3} & \text{if} \frac{7}{20} \leq x \leq \frac{1}{2}, \\ e^{-100x^2} & \text{if} \frac{1}{2} \leq x \leq \frac{3}{2}, \\ 0 & \text{if} \frac{3}{2} \leq x \leq 2, \end{cases}.$$

Example 7 In the seven example, we consider a non-smooth source function as follows

$$f(x) = \begin{cases} 0 & \text{if} 0 \leq x \leq \frac{2}{5}, \\ -3x + 9 & \text{if} \frac{2}{5} \leq x \leq \frac{4}{5}, \\ 6 - \sqrt{1 - x^2} & \text{if} \frac{4}{5} \leq x \leq \frac{6}{5}, \\ 3x + 9 & \text{if} \frac{6}{5} \leq x \leq \frac{8}{5}, \\ 0 & \text{if} \frac{8}{5} \leq x \leq 2 \end{cases}.$$

Example 8 In the eight example, we consider a non-smooth source function as follows

$$f(x) = \frac{1}{10}|x - 1|e^{-5|x-1|}.$$

The data ϕ is obtained by solving the direct problem (1).

6.1.1 Reconstruction Results Without Noise

In this paragraph we present the numerical result of source term in one-dimensional case without noise for the examples above and the relative errors for each of these examples. We start with the smooth examples like 1 and 2 where the result are presented in Fig. 1. While the results of Examples 3 and 4 are displayed in Fig. 2. Figure 3 shows the estimation results for both nonsmooth Examples 5 and 6. In Fig. 4, we present the results for both nonsmooth Examples 7 and 8.

It turns out that, the results obtained in these smooth examples almost coincide with the exact solution. On the other hand, for the nonsmooth examples, the method does not build exactly the singularities of the sources but approaches them.

The relative error for Examples 1 and 2 are are shown in Fig. 5. While, for Examples 3 and 4 are displayed in Fig. 6. In Figs. 7 and 8, we present The relative error for non-smooth Examples 5–6 and 7–8, respectively. Which shows that the convergence behaviour of Algorithm 1.

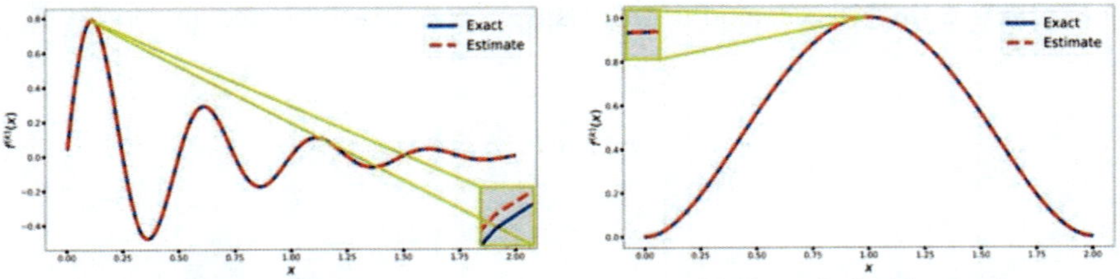

(a) Example 1 without noise (b) Example 2 without noise

Fig. 1 Numerical result of the exact solution $f(x)$ and the estimate solution $f^{(k)}(x)$ for Examples 1 and 2 without noise

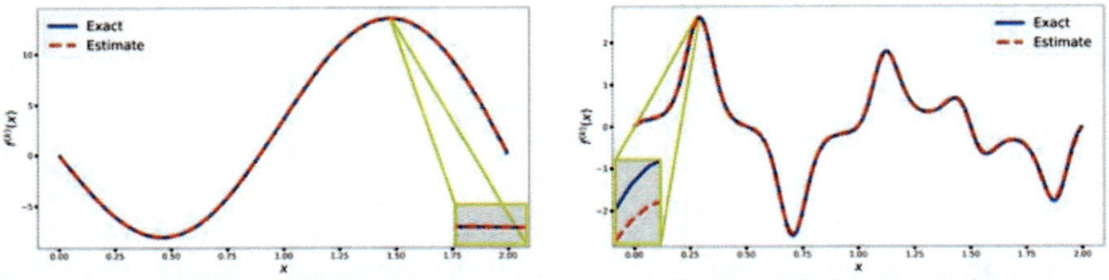

(a) Example 3 without noise (b) Example 4 without noise

Fig. 2 Numerical result of the exact solution $f(x)$ and the estimate solution $f^{(k)}(x)$ for Examples 3 and 4 without noise

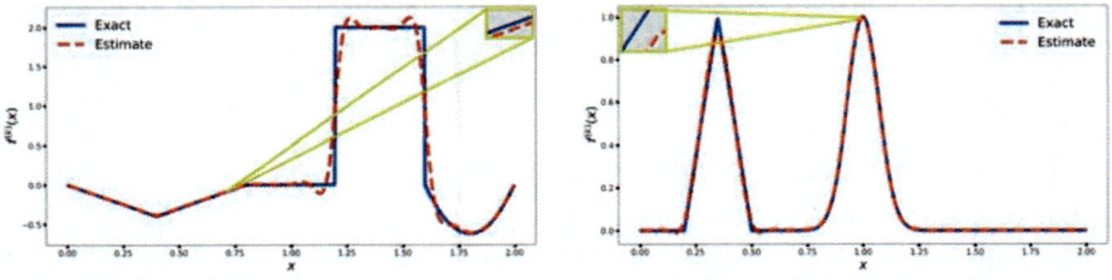

(a) Example 5 with noise (b) Example 6 without noise

Fig. 3 Numerical result of the exact solution $f(x)$ and the estimate solution $f^{(k)}(x)$ for Examples 5 and 6 with noise

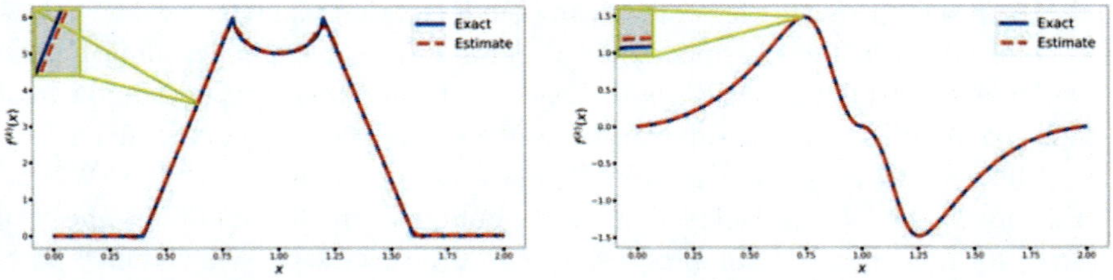

(a) Example 7 with noise. (b) Example 8 with noise.

Fig. 4 Numerical result of the exact solution $f(x)$ and the estimate solution $f^{(k)}(x)$ for Examples 7 and 8 without noise

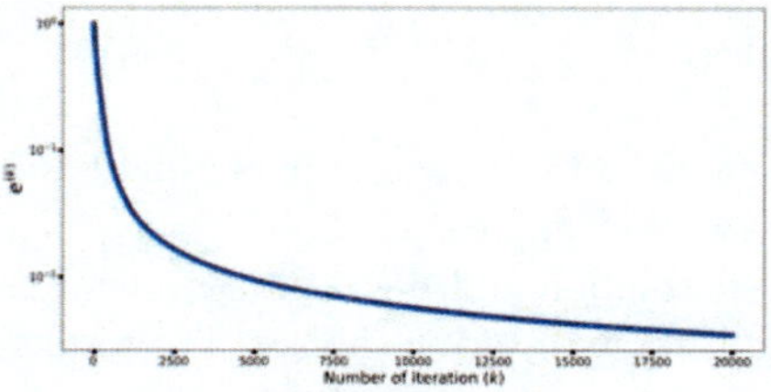

(a) Example 1.

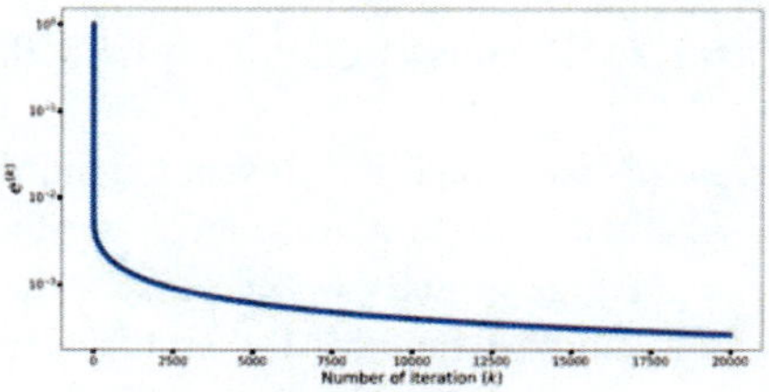

(b) Example 2.

Fig. 5 The relative error of $f(x)$ for Examples 1 and 2

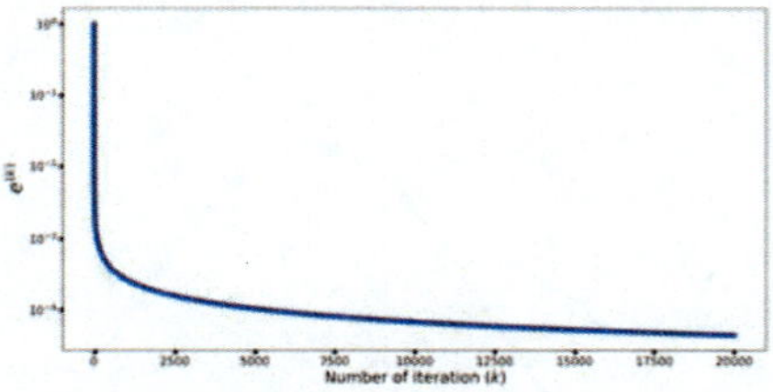

(a) Example 3

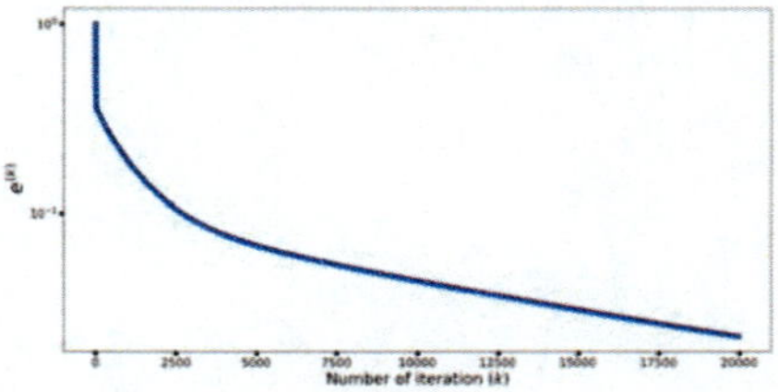

(b) Example 4

Fig. 6 The relative error of $f(x)$ for Examples 3 and 4

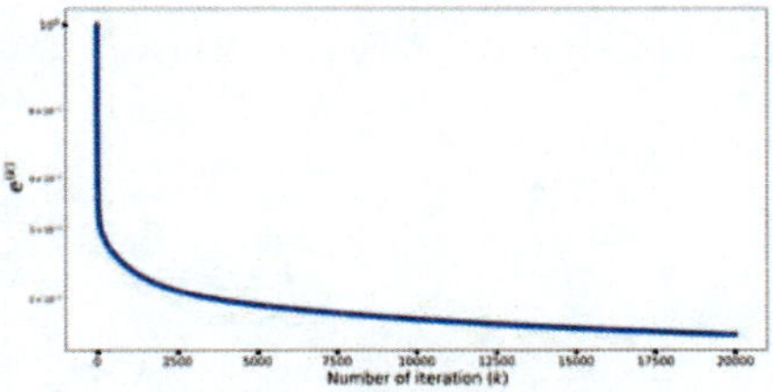

(a) Example5.

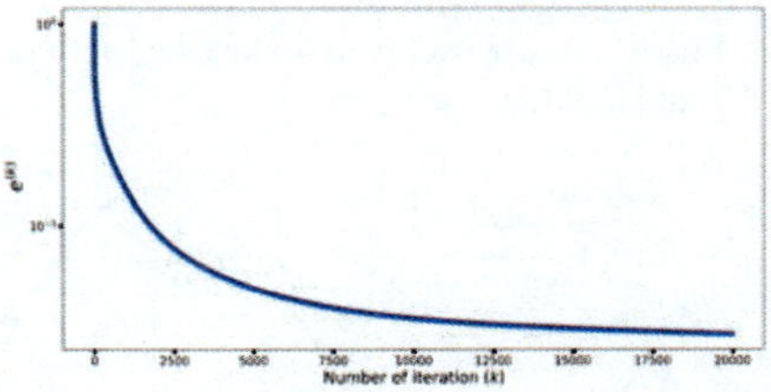

(b) Example 6.

Fig. 7 The relative error of $f(x)$ for Examples 5 and 6

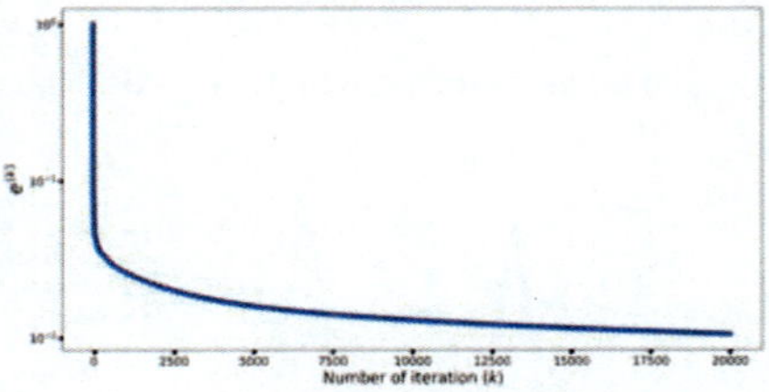

(a) Example 7.

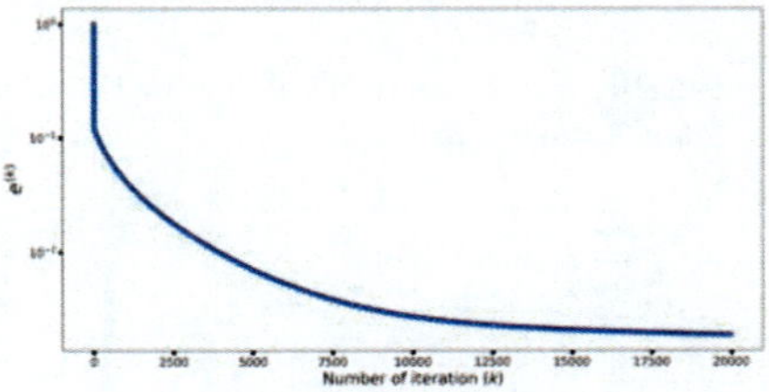

(b) Example 8.

Fig. 8 The relative error of $f(x)$ for Examples 7 and 8

6.1.2 Reconstruction Results with Noise

In this paragraph we present the numerical result of source term in one-dimensional case with different noise levels for the previous examples.

In Figs. 9 and 10, we present the exact solution and the estimate source term for smooth Examples 1–2 and 3–4, respectively with various noises. The results for nonsmooth Examples 5–6 and 7–8 with different noise are displayed in Figs. 11 and 12, respectively.

We observe that the conjugate gradient method is stable in the presence of noise, according to these numerical experiments.

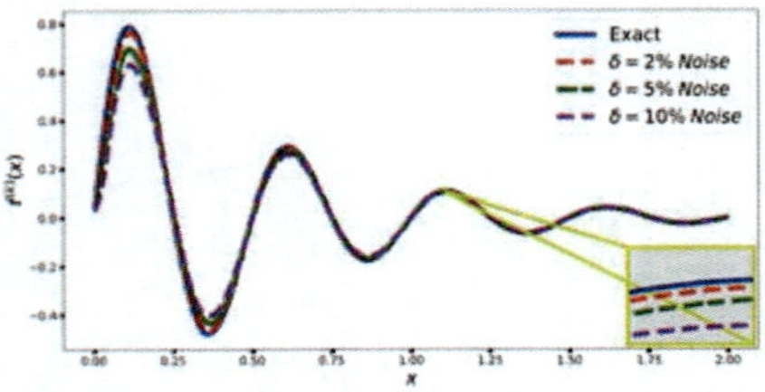

(a) Example 1.

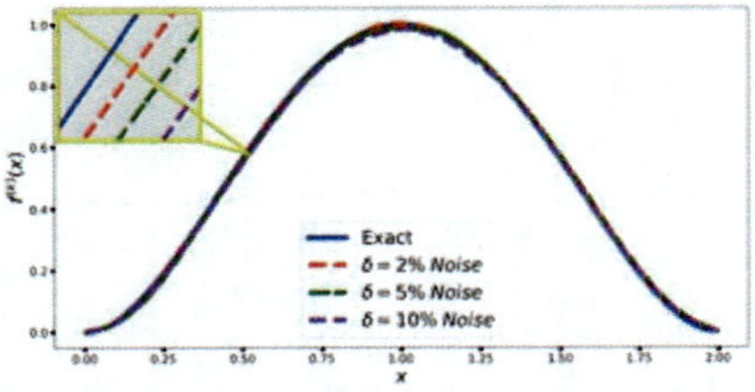

(b) Example 2.

Fig. 9 Numerical result of the exact solution $f(x)$ and the estimate solution $f^{(k)}(x)$ for Examples 1 and 2 with noise

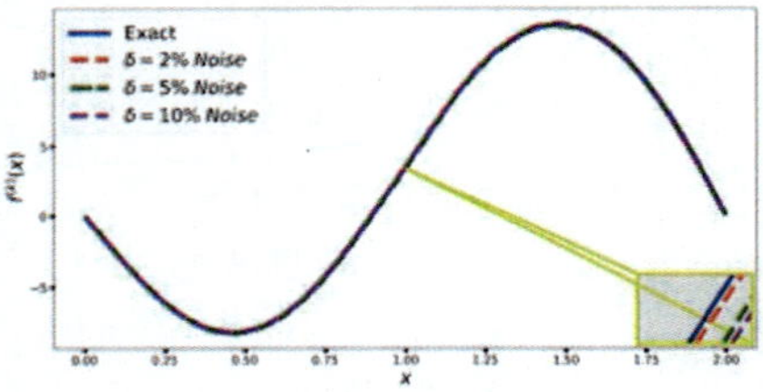

(a) Example 3.

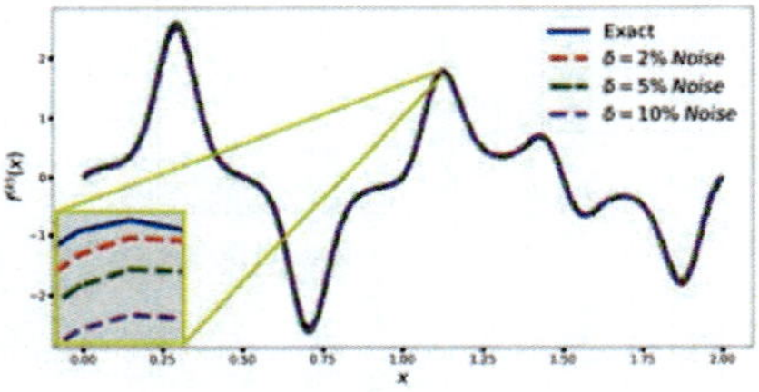

(b) Example 4.

Fig. 10 Numerical result of the exact solution $f(x)$ and the estimate solution $f^{(k)}(x)$ for Examples 3 and 4 with noise

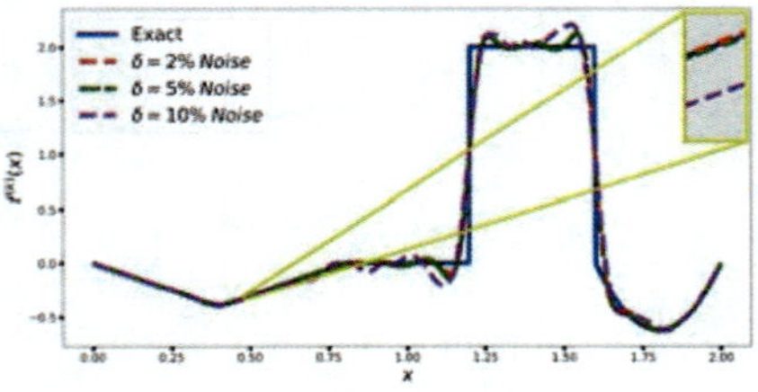

(a) Example 5.

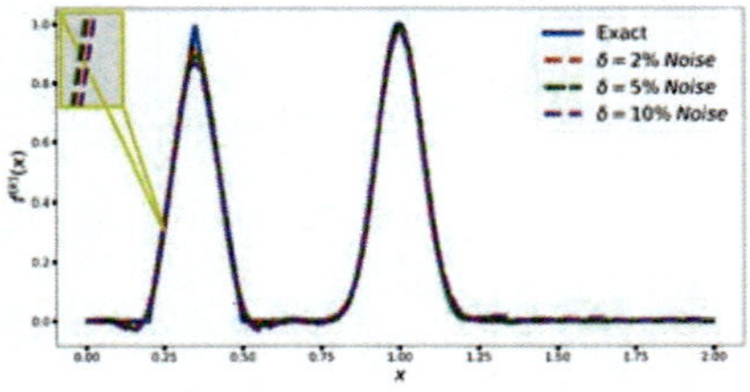

(b) Example 6.

Fig. 11 Numerical result of the exact solution $f(x)$ and the estimate solution $f^{(k)}(x)$ for Examples 5 and 6 with noise

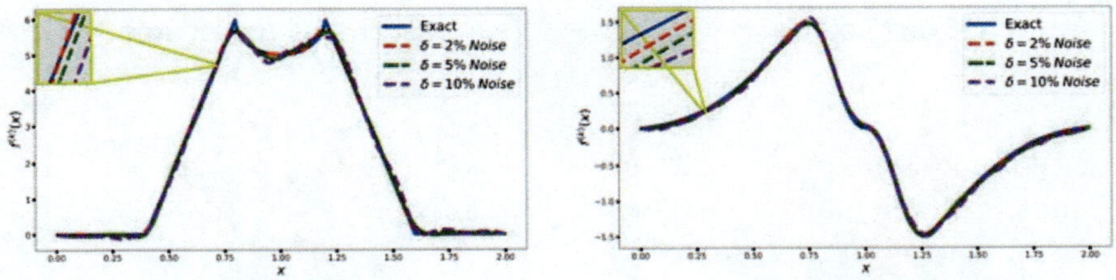

(a) Example 7. (b) Example 8.

Fig. 12 Numerical result of the exact solution $f(x)$ and the estimate solution $f^{(k)}(x)$ for Examples 7 and 8 with noise

6.2 *Results for the Two-Dimensional Case*

In the subsection, we present four examples in two-dimensional case. Among these four examples, we take the first three examples are smooth and the last example are nonsmooth. In the numerical computation, we take $n = 2$, $\boldsymbol{\Omega} = (0, 2) \times (0, 2)$ and $\lambda = 10^{-5}$.

Example 9 In the example, we consider a smooth source function defined as follows

$$f(x, y) = \sin(2\pi x)\sin(2\pi y).$$

Example 10 In the example, we consider a smooth source function as follows

$$f(x, y) = (1 - \cos(2\pi x)(1 - \cos(2\pi y) + \frac{3}{5}.$$

Example 11 In the example, we consider a smooth source function as follows

$$f(x, y) = e^{x+y}x(2 - x)y(2 - y) + \frac{5}{2}.$$

Example 12 In the example, we consider a non-smooth source function as follows

$$f(x, y) = (1 - e^{x^2-2x|x-1|})\sin(4\pi|y - \frac{1}{2}|) + \frac{1}{2}.$$

The data ϕ is obtained by solving the direct problem (1).

6.2.1 Reconstruction Results Without Noise

In this paragraph we present the numerical result of source term in two-dimensional case without noise for the four examples above.

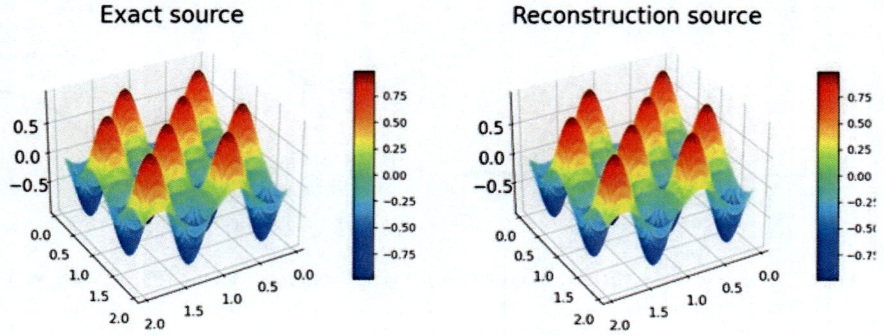

Fig. 13 Exact solution $f(x, y)$ and estimate solution $f^{(k)}(x, y)$ for Example 9

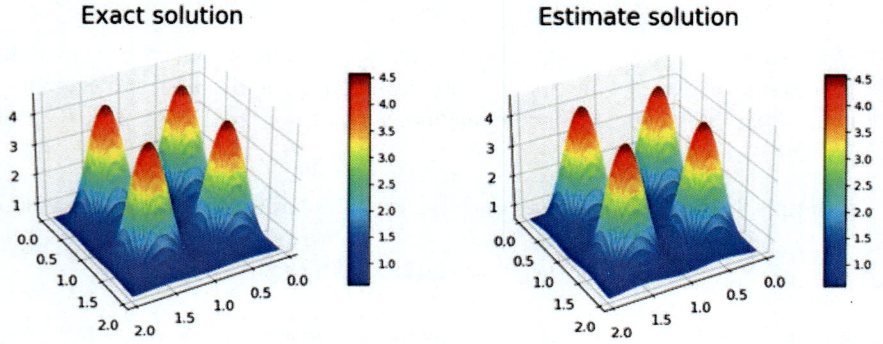

Fig. 14 Exact solution $f(x, y)$ and estimate solution $f^{(k)}(x, y)$ for Example 10

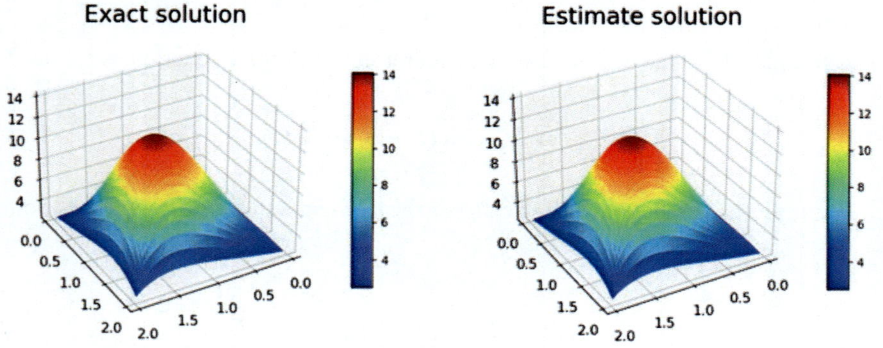

Fig. 15 Exact solution $f(x, y)$ and estimate solution $f^{(k)}(x, y)$ for Example 11

We present in Figs. 13–15 the estimate solution compared to the exact solution without noise for smooth Examples 9, 10 and 11, respectively. While, the result of nonsmooth Example 12 is shown in Fig. 16.

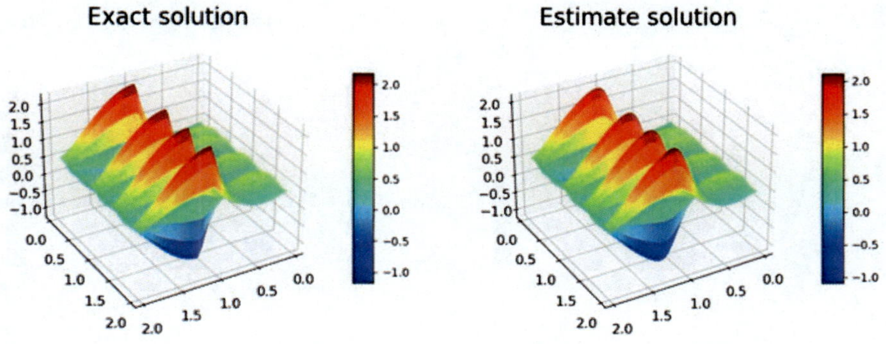

Fig. 16 Exact solution $f(x, y)$ and estimate solution $f^{(k)}(x, y)$ for Example 12

6.2.2 Reconstruction Results With Noise

In this paragraph we present the numerical result of source term in two-dimensional case with different noise levels for the four previous examples above. In Figs. 17–19, we show the exact solution and the estimate source with different levels of noise for smooth Examples 9, 10 and 11, respectively. While, the numerical test of nonsmooth Example 12 is presented in Fig. 20.

All the tests done before whether in one-dimensional or two-dimensional space show the stability of our proposed approach.

7 Conclusion

In this work, we have presented a study on the inverse source problem of estimating a source term in the linear elliptic equation from observed measurements. We started this study by showing that the direct problem is well-posed, formulated our inverse source problem into a control problem using the Tikhonov regularization method, then established the existence and uniqueness of the minimizer. The conjugate gradient algorithm is utilized to solve the regularized control problem, and some numerical examples are provided to show that the conjugate gradient method is effective and stable for solving the inverse source problem.

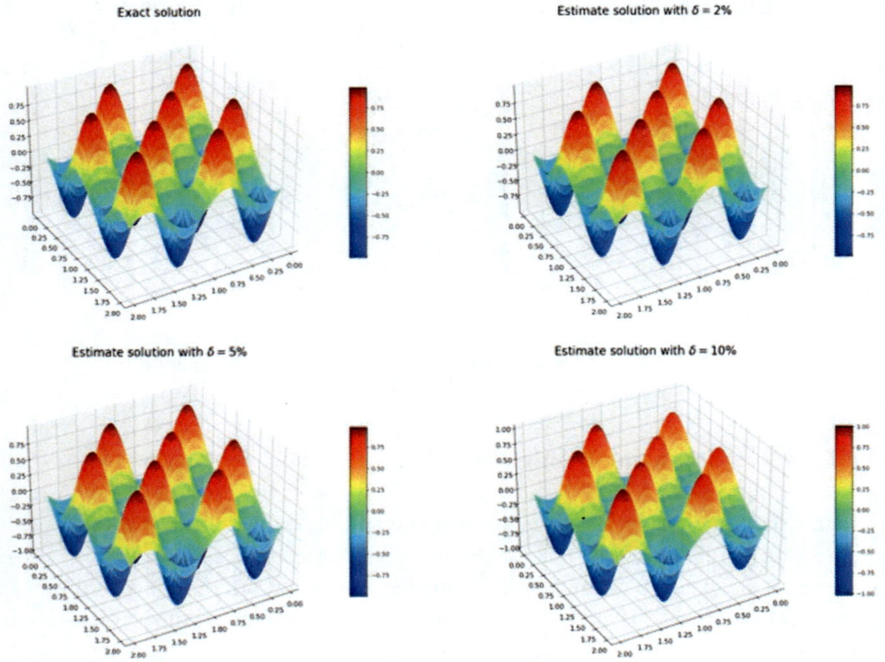

Fig. 17 Numerical result of the exact solution $f(x, y)$ and the estimate solution $f^{(k)}(x, y)$ for Example 9 with noise

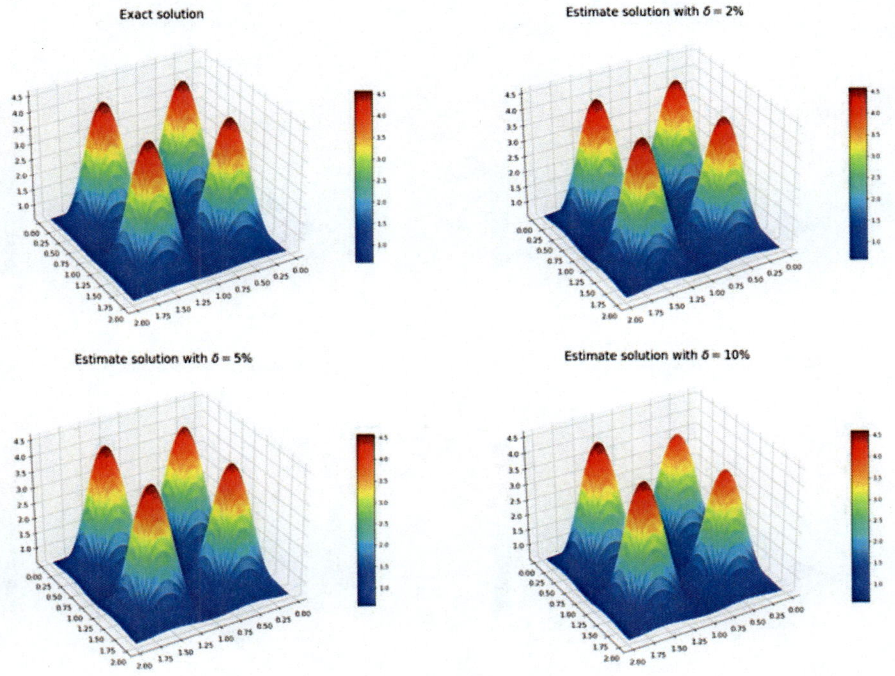

Fig. 18 Numerical result of the exact solution $f(x, y)$ and the estimate solution $f^{(k)}(x, y)$ for Example 10 with noise

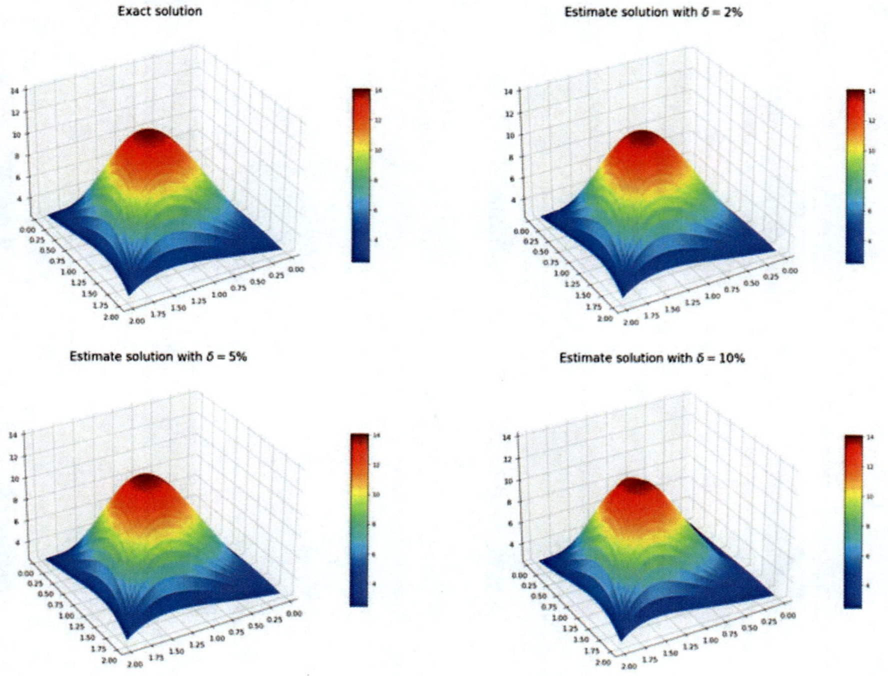

Fig. 19 Numerical result of the exact solution $f(x, y)$ and the estimate solution $f^{(k)}(x, y)$ for Example 11 with noise

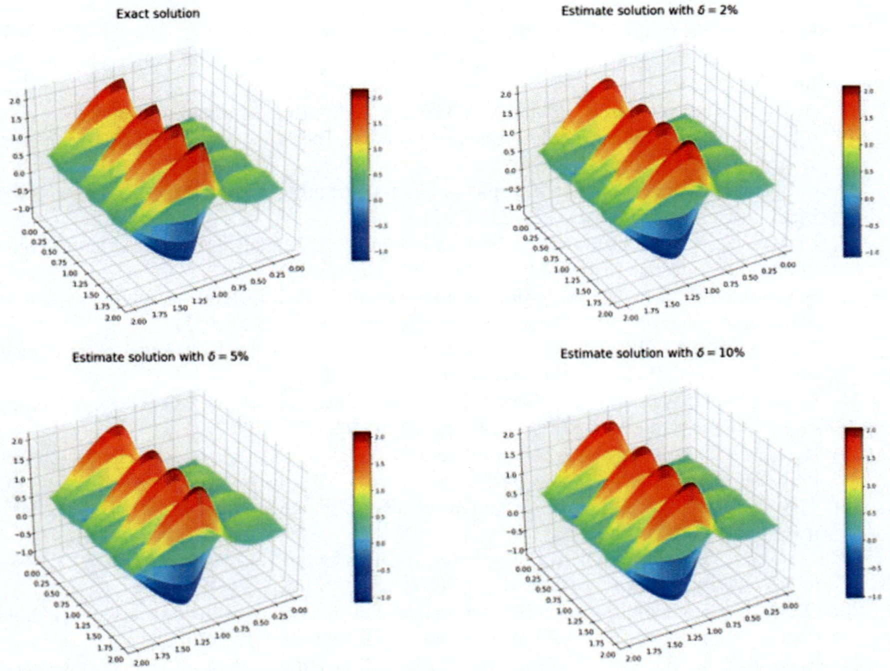

Fig. 20 Numerical result of the exact solution $f(x, y)$ and the estimate solution $f^{(k)}(x, y)$ for Example 12 with noise

References

1. V. Isakov, *Inverse Source Problems*, vol. 34. (American Mathematical Soc., 1990)
2. L. Afraites, A. Hadri, A. Laghrib, M. Nachaoui, A weighted parameter identification pde-constrained optimization for inverse image denoising problem. Vis. Comput. **38**(8), 2883–2898 (2022)
3. A. Oulmelk, L. Afraites, A. Hadri, An inverse problem of identifying the coefficient in a nonlinear time-fractional diffusion equation. Comput. Appl. Math. **42**(1), 65 (2023)
4. M. Hämäläinen, R. Hari, R.J. Ilmoniemi, J. Knuutila, O.V. Lounasmaa, Magnetoencephalography–theory, instrumentation, and applications to noninvasive studies of the working human brain. Rev. Mod. Phys. **65**(2), 413 (1993)
5. S. Lyaqini, M. Nachaoui, Identification of genuine from fake banknotes using an enhanced machine learning approach, in *International Conference on Numerical Analysis and Optimization Days* (Springer, 2021), pp. 59–70
6. H. Werner Engl, M. Hanke, A. Neubauer, *Regularization of Inverse Problems*, vol. 375. (Springer Science & Business Media, 1996)
7. A. Laghrib, L. Afraites, A. Hadri, M. Nachaoui, *A non-convex pde-constrained denoising model for impulse and gaussian noise mixture reduction* (Inverse Prob, Imaging, 2022)
8. A. El Badia, T.H. Duong, Some remarks on the problem of source identification from boundary measurements. Inverse Prob. **14**(4), 883 (1998)
9. M. Nachaoui, L. Afraites, A. Hadri, A. Laghrib, A non-convex non-smooth bi-level parameter learning for impulse and gaussian noise mixture removing. Commun. Pure Appl. Anal. **21**(4), 1249 (2022)

10. L. Afraites, A. Hadri, A. Laghrib, M. Nachaoui, *A non-convex denoising model for impulse and gaussian noise mixture removing using bi-level parameter identification* (Inverse Prob, Imaging, 2022)
11. A. Hadri, M. Nachaoui, A. Laghrib, A. Chakib, L. Afraites, A primal-dual approach for the robin inverse problem in a nonlinear elliptic equation: The case of the $L^1 - L^2$ cost functional. J. Inverse Ill-Posed Prob. (2022)
12. S. Lyaqini, M. Nachaoui, Diabetes prediction using an improved machine learning approach. Math. Model. Comput. **8**(1), 726–735 (2021)
13. L. Afraites, A. Atlas, Parameters identification in the mathematical model of immune competition cells. J. Inverse Ill-Posed Prob. **23**(4), 323–337 (2015)
14. S. Lyaqini, M. Nachaoui, A. Hadri, An efficient primal-dual method for solving non-smooth machine learning problem. Chaos Solitons Fractals **155**, 111754 (2022)
15. C.J.S. Alves, J.B. Abdallah, M. Jaoua, Recovery of cracks using a point-source reciprocity gap function. Inverse Prob. Sci. Eng. **12**(5), 519–534 (2004)
16. L. Afraites, C. Masnaoui, M. Nachaoui, Coupled complex boundary method for a geometric inverse source problem. RAIRO-Oper. Res. **56**(5), 3689–3709 (2022)
17. L. Ling, Y.C. Hon, M. Yamamoto, Inverse source identification for poisson equation. Inverse Prob. Sci. Eng. **13**(4), 433–447 (2005)
18. T. Nara, S. Ando, A projective method for an inverse source problem of the poisson equation. Inverse Prob. **19**(2), 355 (2003)
19. T. Shigeta, Y.-C. Hon, Numerical source identification for poisson equation, in *Inverse Problems in Engineering Mechanics IV* (Elsevier, 2003), pp. 137–145
20. O. Xie, Z. Zhao, Identifying an unknown source in the poisson equation by a modified tikhonov regularization method. Int. J. Math. Comput. Sci. **6**(8), 1128–1132 (2012)
21. L. Afraites, C. Masnaoui, M. Nachaoui, Shape optimization method for an inverse geometric source problem and stability at critical shape. Dis. Cont. Dyn. Syst.-S **15**(1), 1 (2022)
22. F. Yang, The truncation method for identifying an unknown source in the poisson equation. Appl. Math. Comput. **217**(22), 9334–9339 (2011)
23. F. Yang, F. Chu-Li, The modified regularization method for identifying the unknown source on poisson equation. Appl. Math. Model. **36**(2), 756–763 (2012)
24. L. Afraites, A. Oulmelk, Identification of robin coefficient in elliptic problem by a coupled complex boundary method, in *International Conference on Numerical Analysis and Optimization Days* (Springer, 2021), pp. 71–86
25. Z. Zhao, Z. Meng, L. You, O. Xie, Identifying an unknown source in the poisson equation by the method of tikhonov regularization in hilbert scales. Appl. Math. Model. **38**(19–20), 4686–4693 (2014)
26. L. Afraites, A new coupled complex boundary method (ccbm) for an inverse obstacle problem. Dis. Cont. Dyn. Syst.-S **15**(1), 23 (2022)
27. J. Cheng, V. Isakov, L. Shuai, Increasing stability in the inverse source problem with many frequencies. J. Differ. Equ. **260**(5), 4786–4804 (2016)
28. C. Ashyralyyev, M. Dedeturk, Approximate solution of inverse problem for elliptic equation with overdetermination, in *Abstract and Applied Analysis*, vol. 2013 (Hindawi, 2013)
29. D.G. Orlovskii, Inverse dirichlet problem for an equation of elliptic type. Differ. Equ. **44**(1), 124–134 (2008)
30. I. Knowles, Parameter identification for elliptic problems. J. Comput. Appl. Math. **131**(1–2), 175–194 (2001)
31. M.S. Gockenbach, The output least-squares approach to estimating lamé moduli. Inverse Prob. **23**(6), 2437 (2007)
32. L.P. Lebedev, I.I. Vorovich, G. Maurice Leslie Gladwell, *Functional Analysis: Applications in Mechanics and Inverse Problems*, vol. 41 (Springer Science & Business Media, 2012)
33. M.F. Al-Jamal, M.S. Gockenbach, Stability and error estimates for an equation error method for elliptic equations. Inverse Prob. **28**(9), 095006 (2012)
34. J.M. Krilow, *Distant Drumming: Morphological Correlates of Habitat and Courtship Behaviour in the Ruffed Grouse (Bonasa umbellus)* (University of Lethbridge, Canada, 2014)

35. A. Oulmelk, L. Afraites, A. Hadri, M. Nachaoui, An optimal control approach for determining the source term in fractional diffusion equation by different cost functionals. Appl. Numer. Math. **181**, 647–664 (2022)
36. A. Oulmelk, M. Srati, L. Afraites, A. Hadri, Implementation of the ADMM approach to constrained optimal control problem with a nonlinear time-fractional diffusion equation. Dis. Cont. Dyn. Syst.-S (2022)
37. X. Bin Yan, T. Wei, Inverse space-dependent source problem for a time-fractional diffusion equation by an adjoint problem approach. J. Inverse Ill-Posed Prob. **27**(1), 1–16 (2019)
38. N. Mourad, L. Afraites, L. Amine, A regularization by denoising super-resolution method based on genetic algorithms. Sig. Proces. Image Commun. **99**, 116505 (2021)
39. L. Afraites, H. Aissam, L. Amine, N. Mourad, A high order PDE-constrained optimization for the image denoising problem. Inverse Prob. Sci. Eng. **29**(12), 1821–1863 (2021)
40. G.S. Abdoulaev, K. Ren, A.H. Hielscher, Optical tomography as a pde-constrained optimization problem. Inverse Prob. **21**(5), 1507 (2005)
41. M.R. Garvie, P.K. Maini, C. Trenchea, An efficient and robust numerical algorithm for estimating parameters in turing systems. J. Comput. Phys. **229**(19), 7058–7071 (2010)
42. application to ultrasound images, H. Aissam. L. Afraites, L. Amine, N. Mourad, A novel image denoising approach based on a non-convex constrained PDE. Sig. Image Video Proces. **15**, 1057–1064 (2021)
43. L. Afraites, A. Hadri, A. Laghrib, M. Nachaoui, A non-convex denoising model for impulse and Gaussian noise mixture removing using bi-level parameter identification. Inverse Prob. Imaging **16**(4), 827–870 (2022)
44. H. Badry, L. Oufni, H. Ouabi, H. Iwase, L. Afraites, A new fast algorithm to achieve the dose uniformity around high dose rate brachytherapy stepping source using Tikhonov regularization. Australas. Phys. Eng. Sci. Med. **42**, 757–769 (2019)

A Mesh Free Wavelet Method to Solve the Cauchy Problem for the Helmholtz Equation

Abdeljalil Nachaoui and Sudad Musa Rashid

Abstract In this paper, we present a numerical based on Haar wavelets to solve an inverse Cauchy problem governed by the Helmholtz equation. The problem involves reconstructing the boundary condition on an inaccessible boundary from the given Cauchy data on another part of the boundary. We discuss the formulation of the problem and the use of Haar wavelets. The proposed method involves approximating the solution using a finite sum of Haar wavelets and solving the resulting linear system of equations using a least-squares method. The effectiveness of the proposed method is demonstrated through numerical experiments. From these results, we demonstrate that the Haar wavelet method can be used to obtain an accurate solution to the problem.

Keywords Inverse problems · Ill-posed problems · Cauchy problem · Helmholtz equation · Meshless method · Haar wavelet method · Multi scale preconditioning

1 Introduction

Inverse problems are ubiquitous in many fields of science and engineering, and their solution plays an important role in understanding and controlling various natural and artificial systems [1–25]. Inverse problems are a class of problems in which one seeks to determine the unknown properties of a physical system (the unknown quantity) from indirect measurements of its response (the observed data) [16, 21, 22, 26–38]. In particular, inverse Cauchy problems are those where the boundary condition of a partial differential equation (PDE) needs to be determined from the knowledge of the solution on another part of the boundary. Solving inverse problems is a chal-

A. Nachaoui (✉)
Laboratoire de Mathématiques Jean Leray, Nantes Université, Nantes, France
e-mail: Abdeljalil.Nachaoui@univ-nantes.fr

S. M. Rashid
Department of Mathematics, College of Basic Education, Sulaimani university, Sulaimani, Iraq
e-mail: sudad.rasheed@univsul.edu.iq

A. Laghrib et al. (eds.), *New Trends of Mathematical Inverse Problems and Applications*, Springer Proceedings in Mathematics & Statistics 428,
https://doi.org/10.1007/978-3-031-33069-8_6

lenging task due to several factors, including ill-posedness, under-determination, and instability. Ill-posedness refers to the fact that small errors in the data or in the solution can lead to large errors in the computed parameters or conditions [8, 39, 40]. Under-determination refers to the fact that the available data may not provide enough information to uniquely determine the parameters or conditions. Instability refers to the fact that the computed solution may be sensitive to small perturbations in the input data or in the numerical method used to solve the problem. These challenges require the use of specialized numerical methods and regularization techniques to obtain accurate and reliable solutions.

Inverse problems can be solved using various numerical methods, such as regularization methods, optimization methods, Bayesian methods, and numerical discretization methods. The choice of method depends on the specific properties of the problem, such as the type of data, the regularity of the solution, and the computational resources available.

Such problems are ill-posed, meaning that small errors in the data can lead to large errors in the solution. Hence, the development of accurate and robust numerical methods for solving inverse Cauchy problems is an important area of research.

In this paper, we consider an inverse Cauchy problem governed by the Helmholtz equation. Specifically, we seek to determine the boundary condition on an inaccessible boundary from the given data on another part of the boundary. Such problems arise in many applications, such as tomography, imaging, geophysics, and medical diagnosis.

With regard to solving the Cauchy problem for the Helmholtz equation, several methods have been proposed over the past two decades [3, 5, 14, 36, 41–49].

In recent years, there has been significant interest in the use of wavelet methods for solving partial differential equations [50–60], as wavelet methods can capture singularities and discontinuities in the solution. In this paper, we investigate the use of Haar wavelets to solve an inverse Cauchy problem governed by the Helmholtz equation. We will follow the approach used in [61].

The rest of the paper is arranged as follows. In Sect. 2, we review the the Helmholtz equation, the inverse Cauchy problem of reconstructing the boundary condition on an accessible boundary from the given data on another part of the boundary and Haar wavelets with related notations. In Sect. 3 we follow the idea proposed in [61], and extend it to the Cauchy problem for the Helmholtz equation. Since in many situations the exact data are not available and one has only noisy measured data, we will consider cases where the given data are computed as a solution of a direct problem. Thus we developed in Sect. 4 formulas for solving this kind of problems. Since the inverse Cauchy problem is ill-posed, linear systems of equations obtained by applying formulas (developed in Sects. 3 and 4) on collocation points in the interior of the domain are ill-conditioned. We briefly review, in Sect. 5 the regularization and preconditioning techniques used to obtain accurate solutions. In Sect. 6, we demonstrate the effectiveness of our method through numerical experiments.

2 Statement of the Inverse Cauchy Problem

2.1 Brief Overview of the Helmholtz Equation

The Helmholtz equation is a partial differential equation that arises in many areas of physics, engineering, and mathematics. It is a second-order linear differential equation that can be used to model wave propagation, such as sound and light waves. The Helmholtz equation takes the form:

$$\nabla^2 u + k^2 u = f, \tag{1}$$

where u is the unknown function, f is a given source term, and k is a constant known as the wave number. The Laplacian operator ∇^2 is defined as the divergence of the gradient of a function.

The Helmholtz equation has physical meaning in many applications. For example, in acoustics, it describes the propagation of sound waves in a homogeneous medium, where k is proportional to the frequency of the sound waves. In electromagnetic, it describes the propagation of electromagnetic waves, such as light, in a medium. In fluid dynamics, it describes the behaviour of fluid waves, such as water waves or air waves. The wave number k plays a crucial role in determining the behaviour of the solution, with higher values of k corresponding to shorter wavelengths and faster oscillations.

2.2 The Inverse Problem

The inverse Cauchy problem is a class of the inverse problems where one seeks to determine the boundary conditions of a PDE from partial data of the solution on a part of the boundary. In other words, given partial data of the solution, the goal is to determine the boundary conditions that led to the observed data.

Let $\Omega = (0, B) \times (0, C)$ with boundary $\partial\Omega = \Gamma_0 \cup \Gamma_1 \cup \Gamma_2 \cup \Gamma_3$ (Fig. 1). Where $\Gamma_0 = \{(x, y) : y = 0\}$, $\Gamma_1 = \{(x, y) : x = 0\}$, $\Gamma_2 = \{(x, y) : x = B\}$ and $\Gamma_3 = \{(x, y) : y = C\}$.

We consider the following inverse Cauchy problem . Find u satisfying the Helmholtz equation (1) in Ω and the following boundary conditions:

$$u(x, 0) = \omega_0(x) \qquad (x, y) \in \Gamma_0, \tag{2}$$

$$u(0, y) = \omega_1(y) \qquad (x, y) \in \Gamma_1, \tag{3}$$

$$u(B, y) = \omega_2(y) \qquad (x, y) \in \Gamma_2, \tag{4}$$

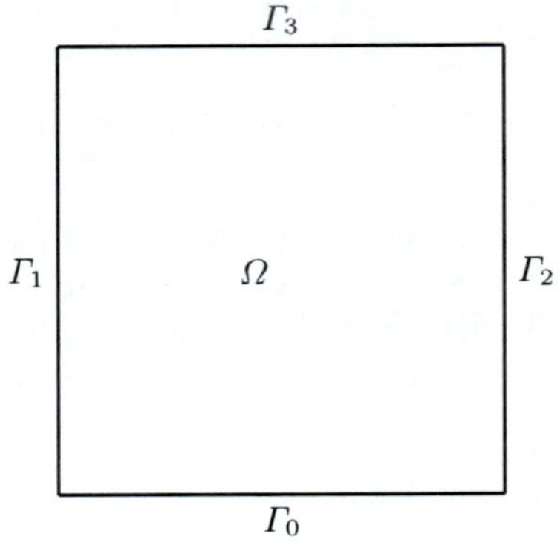

Fig. 1 Geometry domain of inverse Cauchy problem

$$\frac{\partial}{\partial y}u(x,0) = g(x) \qquad (x, y) \in \Gamma_0. \tag{5}$$

The Inverse problem (1)–(5) is ill-posed, which means that its solution is sensitive to noise or perturbations in the data. Therefore, numerical methods play a crucial role in solving this kind of inverse problems by providing stable, accurate, and efficient solutions. It can be solved using various numerical methods, such as regularization methods, optimization methods, Bayesian methods, and numerical discretization methods. The choice of method depends on the specific properties of the problem, such as the type of data, the regularity of the solution, and the computational resources available. In the following we developed a method based on Haar wavelets.

2.3 *Overview of the Haar Wavelet Method*

The Haar wavelet method is a numerical discretization method that uses the Haar wavelet basis functions to approximate the solution of a partial differential equation. The Haar wavelets are a set of piecewise constant functions that have a compact support and an orthogonal property. The Haar wavelet method has several advantages over other numerical methods, such as simplicity of implementation.

Wavelet-based numerical methods have been widely applied in differential equation [51, 53–55] and various types of PDEs [50, 56–58, 60]. Mathematically, the simplest of all these families of wavelets are the Haar wavelets [62]. They consist of pairs of piecewise constant functions and can be analytically integrated at arbitrary instants. The general idea of these methods is to develop the Haar series solution then the whole system is discretized by the Galerkin or collocation method. Here, we will follow an idea recommended in [50], namely to develop in the Haar series not the function, but its highest derivative appearing in the differential equation. We then obtain by integration, the other derivatives as well as the function itself. We will then use the boundary conditions in order to express the solution of the problem with a semi-analytical formula whose coefficients can be obtained after discretization by collocation.

Let us define $M = 2^J$, where J is the maximal level of resolution. Then let us introduce the dilatation parameter (or the level of resolution) $j = 0, 1, 2, \ldots, J$ and the translation parameter $k = 0, 1, 2, \ldots, m-1$ with $m = 2^j$.

For $x \in [a, b]$, the orthogonal set of Haar wavelet functions are defined by:

$$h_1(x) = \begin{cases} 1 & \text{for } x \in [a, b), \\ 0 & \text{otherwise}, \end{cases} \tag{6}$$

$$h_2(x) = \begin{cases} 1 & \text{for } x \in [a, \frac{a+b}{2}), \\ -1 & \text{for } x \in [\frac{a+b}{2}, b), \\ 0 & \text{elsewhere }, \end{cases} \tag{7}$$

$$h_i(x) = \begin{cases} 1 & \text{for } x \in [\xi_1(i), \xi_2(i)), \\ -1 & \text{for } x \in [\xi_2(i), \xi_3(i)), \\ 0 & \text{elsewhere }, \end{cases} \tag{8}$$

where the index i in $h_i(x)$ is computed by $i = m + k + 1$ and

$$\xi_1(i) = a + \frac{k(b-a)}{m}, \qquad \xi_2(i) = a + \frac{(k+0.5)(b-a)}{m}, \qquad \xi_3(i) = a + \frac{(k+1)(b-a)}{m}.$$

This writing supposes that the interval $[a, b]$ is divided into $2M$ sub-intervals of length equal to $\Delta x = \frac{b-a}{2M}$

2.4 *Function Approximation*

A function $u(x) \in L^2([a, b])$ can be expanded into Haar wavelet by

$$u(x) = \sum_{i=1}^{\infty} u_i h_i(x) \tag{9}$$

where $u_i = \int_a^b u(x)h_i(x)dx$. If $u(x)$ is approximated as piecewise constant during each subinterval, Eq. (9) will be terminated at finite terms

$$u(x) = \sum_{i=1}^{m_t} u_i h_i(x) \tag{10}$$

where m_t is large enough.

A function $u(x, y) \in L^2([a, b] \times [a, c])$ can be also approximated by Haar wavelet in the form

$$u(x, y) = \sum_{i=1}^{m_t} \sum_{j=1}^{m_t} u_{ij} h_i(x) h_j(y) \tag{11}$$

where

$$u_{ij} = \int_a^c \int_a^b u(x, y) h_i(x) h_j(y) dx dy \tag{12}$$

2.5 *Integrals Haar Wavelets*

Since we want to solve a Cauchy problem governed by Poisson equation which a second order partial differential equation we need the integrate twice Haar wavelets functions thus we need the integrals given for $i \geq 2$ by

$$\begin{aligned} P_{1,i}(x) &= \int_a^x h_i(t) dt, \\ &= \begin{cases} x - \xi_1(i) & \text{for } x \in [\xi_1(i), \xi_2(i)), \\ \xi_3(i) - x & \text{for } x \in [\xi_2(i), \xi_3(i)), \\ 0 & \text{elsewhere }, \end{cases} \end{aligned} \tag{13}$$

and

$$\begin{aligned} P_{2,i}(x) &= \int_a^x P_{1,i}(t) dt, \\ &= \begin{cases} 0 & x < \xi_1(i), \\ \frac{(x - \xi_1(i))^2}{2} & x \in [\xi_1(i), \xi_2(i)), \\ (\xi_3(i) - \xi_2(i))^2 - \frac{(\xi_3(i) - x)^2}{2} & x \in [\xi_2(i), \xi_3(i)), \\ (\xi_3(i) - \xi_2(i))^2 & x \geq \xi_3(i). \end{cases} \end{aligned} \tag{14}$$

In case $i = 1$ we have $\xi_1(1) = a$, $\xi_2(1) = \xi_3(1) = b$ and

$$P_{1,1}(x) = (x - a), \tag{15}$$

$$P_{2,1}(x) = \frac{1}{2}(x - a)^2 \tag{16}$$

The Haar wavelet method can be used to solve various types of partial differential equations, such as elliptic, parabolic, and hyperbolic equations, and their inverse problems. In particular, the Haar wavelet method has been used in [61] to solve inverse Cauchy problems governed by the Poisson equation by discretizing the domain into a set of collocation points and solving the resulting linear system of equations. We follow this idea to solve the inverse problem (1)–(5).

3 Solving Inverse Cauchy Problems Using Haar Wavelets

The idea is to first represent the fourth derivative of the unknown function $u(x, y)$ as a sum of Haar wavelets:

$$\frac{\partial^4 u}{\partial x^2 \partial y^2} = \sum_{i=1}^{2M} \sum_{l=1}^{2M} u_{il} h_i(x) h_l(y), \tag{17}$$

where u_{il} are the wavelet coefficients and $h_i(x)$, $h_l(y)$ the Haar function.

Integrating the Eq. (17) with respect to x from 0 to x, we get

$$\frac{\partial^3 u(x, y)}{\partial x \partial y^2} = \sum_{i=1}^{2M} \sum_{l=1}^{2M} u_{il} P_{1i}(x) h_l(y) + \frac{\partial^2}{\partial y^2}\left[\frac{\partial u(0, y)}{\partial x}\right],$$

Integrating an other time with respect to x from 0 to x, we obtain

$$\frac{\partial^2 u(x, y)}{\partial y^2} = \sum_{i=1}^{2M} \sum_{l=1}^{2M} u_{il} P_{2i}(x) h_l(y) + x \frac{\partial^2}{\partial y^2}\left[\frac{\partial u(0, y)}{\partial x}\right] + \frac{\partial^2 u(0, y)}{\partial y^2}. \tag{18}$$

Putting $x = B$ in the last equation, we get

$$\frac{\partial^2 u(B, y)}{\partial y^2} = \sum_{i=1}^{2M} \sum_{l=1}^{2M} u_{il} P_{2i}(B) h_l(y) + B \frac{\partial^2}{\partial y^2}\left[\frac{\partial u(0, y)}{\partial x}\right] + \frac{\partial^2 u(0, y)}{\partial y^2},$$

from which we find the following expression

$$\frac{\partial^2}{\partial y^2}\left[\frac{\partial u(0, y)}{\partial x}\right] = \frac{1}{B}\frac{\partial^2 u(B, y)}{\partial y^2} - \frac{1}{B}\sum_{i=1}^{2M} \sum_{l=1}^{2M} u_{il} P_{2i}(B) h_l(y) - \frac{1}{B}\frac{\partial^2 u(0, y)}{\partial y^2}. \tag{19}$$

Now, replace (19) in (18), we get

$$\begin{aligned}\frac{\partial^2 u(x,y)}{\partial y^2} &= \sum_{i=1}^{2M}\sum_{l=1}^{2M} u_{il} P_{2i}(x) h_l(y) + \frac{x}{B}\frac{\partial^2 u(B,y)}{\partial y^2} - \frac{x}{B}\frac{\partial^2 u(0,y)}{\partial y^2} \\ &\quad - \frac{x}{B}\sum_{i=1}^{2M}\sum_{l=1}^{2M} u_{il} P_{2i}(B) h_l(y) + \frac{\partial^2 u(0,y)}{\partial y^2}, \\ &= \sum_{i=1}^{2M}\sum_{l=1}^{2M} u_{il}\left\{P_{2i}(x)h_l(y) - \frac{x}{B}P_{2i}(B)h_l(y)\right\} \\ &\quad + \frac{x}{B}\left\{\frac{\partial^2 u(B,y)}{\partial y^2} - \frac{\partial^2 u(0,y)}{\partial y^2}\right\} + \frac{\partial^2 u(0,y)}{\partial y^2} \end{aligned} \tag{20}$$

Integrating the Eq. (20) with regard to y from 0 to y leads to

$$\begin{aligned}\frac{\partial u(x,y)}{\partial y} &= \sum_{i=1}^{2M}\sum_{l=1}^{2M} u_{il}\left\{P_{2i}(x)P_{1l}(y) - \frac{x}{B}P_{2i}(B)P_{1l}(y)\right\} \\ &\quad + \frac{x}{B}\left\{\left(\frac{\partial u(B,y)}{\partial y} - \frac{\partial u(0,y)}{\partial y}\right) - \left(\frac{\partial u(B,0)}{\partial y} - \frac{\partial u(0,0)}{\partial y}\right)\right\} \\ &\quad + \left\{\frac{\partial u(0,y)}{\partial y} - \frac{\partial u(0,0)}{\partial y}\right\} + \frac{\partial u(x,0)}{\partial y}. \end{aligned} \tag{21}$$

Again, integrating (21) with respect to y from 0 to y we get

$$\begin{aligned} u(x,y) &= \sum_{i=1}^{2M}\sum_{l=1}^{2M} u_{il}\left\{P_{2l}(y)\left(P_{2i}(x) - \frac{x}{B}P_{2i}(B)\right)\right\} + y\frac{\partial u(x,0)}{\partial y} + u(x,0) \\ &\quad + \left\{u(0,y) - u(0,0) - y\frac{\partial u(0,0)}{\partial y}\right\} - \frac{xy}{B}\left\{\frac{\partial u(B,0)}{\partial y} - \frac{\partial u(0,0)}{\partial y}\right\} \\ &\quad + \frac{x}{B}\left\{u(B,y) - u(B,0) - u(0,y) + u(0,0)\right\}. \end{aligned} \tag{22}$$

Now, the introduction of the boundary conditions (2)–(5) in Eq. (22) implies that

$$\begin{aligned} u(x,y) &= \sum_{i=1}^{2M}\sum_{l=1}^{2M} u_{il}\left\{P_{2l}(y)\left(P_{2i}(x) - \frac{x}{B}P_{2i}(B)\right)\right\} - \frac{xy}{B}\{g(B) - g(0)\} \\ &\quad + \{\omega_1(y) - \omega_0(0) - yg(0)\} + \frac{x}{B}\{\omega_2(y) - \omega_0(B) + \omega_0(0) - \omega_1(y)\} \\ &\quad + yg(x) + \omega_0(x). \end{aligned} \tag{23}$$

From (23) we obtain Laplace equation as follows

$$\Delta u(x,y)=\sum_{i=1}^{2M}\sum_{l=1}^{2M}u_{il}\left\{h_i(x)P_{2l}(y)+h_l(y)\left(P_{2i}(x)-\frac{x}{B}P_{2i}(B)\right)\right\}+\omega_0^{''}(x)$$
$$+\frac{x}{B}\omega_2^{''}(y)+(1-\frac{x}{B})\omega_1^{''}(y)+yg^{''}(x). \tag{24}$$

Let's introduce Eqs. (23) and (24) in Eq. (1), we obtain

$$f(x,y)=\sum_{i=1}^{2M}\sum_{l=1}^{2M}u_{il}\left\{h_i(x)P_{2l}(y)+h_l(y)\left(P_{2i}(x)-\frac{x}{B}P_{2i}(B)\right)\right\}+\omega_0^{''}(x)$$
$$+\frac{x}{B}\omega_2^{''}(y)+(1-\frac{x}{B})\omega_1^{''}(y)+yg^{''}(x)$$
$$+k^2*\sum_{i=1}^{2M}\sum_{l=1}^{2M}u_{il}\left\{P_{2l}(y)\left(P_{2i}(x)-\frac{x}{B}P_{2i}(B)\right)\right\}+k^2*yg(x)$$
$$-k^2*\frac{xy}{B}\{g(B)-g(0)\}+k^2*\{\omega_1(y)-\omega_0(0)-yg(0)\}$$
$$+k^2*\frac{x}{B}\{\omega_2(y)-\omega_0(B)+\omega_0(0)-\omega_1(y)\}+k^2*\omega_0(x). \tag{25}$$

We write Eq. (25) for any collocation points (x_r, y_s) with $r, s \in \{1, 2, \ldots, 2M\}$ as follow

$$\sum_{i=1}^{2M}\sum_{l=1}^{2M}u_{il}R_{ilrs}=F(r,s), \tag{26}$$

where

$$R_{ilrs}=h_i(x)P_{2l}(y)+\left(P_{2i}(x)-\frac{x}{B}P_{2i}(B)\right)\left(h_l(y)+k^2P_{2l}(y)\right), \tag{27}$$

and

$$F(r,s)=f(x_r,y_s)-\omega_0^{''}(x)-\frac{x}{B}\omega_2^{''}(y)-(1-\frac{x}{B})\omega_1^{''}(y)-yg^{''}(x)-k^2yg(x)$$
$$-k^2*\omega_0(x)-k^2\frac{x}{B}\{\omega_2(y)-\omega_0(B)+\omega_0(0)-\omega_1(y)\}$$
$$+k^2*\frac{xy}{B}\{g(B)-g(0)\}-k^2*\{\omega_1(y)-\omega_0(0)-yg(0)\}. \tag{28}$$

The Eq. (26) expressed in the different collocation points gives rise to a system of linear equations of order $(2M)^2$.

To simulate measured data, we will consider in some numerical examples that $g(x)$ in the condition (5) is obtained as the normal derivative of the solution of a direct problem that we develop in the next section.

4 Haar Wavelets Approximation of the Direct Problem

Consider the following direct problem governed by Helmholtz equation

$$\Delta u(x, y) + k^2 u(x, y) = f(x, y) \qquad (x, y) \in \Omega, \tag{29}$$

$$u(x, 0) = \omega_0(x) \qquad (x, y) \in \Gamma_0, \tag{30}$$

$$u(0, y) = \omega_1(y) \qquad (x, y) \in \Gamma_1, \tag{31}$$

$$u(B, y) = \omega_2(y) \qquad (x, y) \in \Gamma_2, \tag{32}$$

$$u(x, C) = \omega_3(x) \qquad (x, y) \in \Gamma_3. \tag{33}$$

Let us remember that the expression (22) is obtained without the application of any condition and therefore it is valid for the solution of any equation having derivatives of order two in x and in y. It can therefore be used for the direct problem as well as for the inverse problem. So let's replace condition (33) in Eq. (22), we obtain

$$\begin{aligned} u(x, C) = {} & \sum_{i=1}^{2M}\sum_{l=1}^{2M} u_{il} \left\{ P_{2l}(C) \left(P_{2i}(x) - \frac{x}{B} P_{2i}(B) \right) \right\} + C \frac{\partial u(x,0)}{\partial y} + u(x, 0) \\ & + \left\{ u(0, C) - u(0, 0) - C \frac{\partial u(0,0)}{\partial y} \right\} - \frac{xC}{B} \left\{ \frac{\partial u(B,0)}{\partial y} - \frac{\partial u(0,0)}{\partial y} \right\} \\ & + \frac{x}{B} \left\{ u(B, C) - u(B, 0) - u(0, C) + u(0, 0) \right\}, \end{aligned} \tag{34}$$

Introducing boundary conditions (30)–(32) in this last equation implies that

$$\begin{aligned} \frac{\partial u(x,0)}{\partial y} = {} & \frac{1}{C}\omega_3(x) - \frac{1}{C} \sum_{i=1}^{2M}\sum_{l=1}^{2M} u_{il} \left\{ P_{2l}(C) \left(P_{2i}(x) - \frac{x}{B} P_{2i}(B) \right) \right\} - \frac{1}{C}\omega_0(x) \\ & - \frac{1}{C} \left\{ \omega_1(C) - \omega_0(0) - C \frac{\partial u(0,0)}{\partial y} \right\} + \frac{x}{B} \left\{ \frac{\partial u(B,0)}{\partial y} - \frac{\partial u(0,0)}{\partial y} \right\} \\ & - \frac{x}{BC} \left\{ \omega_2(C) - \omega_0(B) - \omega_1(C) + \omega_0(0) \right\}. \end{aligned} \tag{35}$$

The introduction of (35) in (22) and some manipulations we obtain

$$
\begin{aligned}
u(x, y) = & \sum_{i=1}^{2M} \sum_{l=1}^{2M} u_{il} \left\{ \left(P_{2i}(x) - \frac{x}{B} P_{2i}(B) \right) \left(P_{2l}(y) - \frac{y}{C} P_{2l}(C) \right) \right\} \\
& + (1 - \frac{y}{C}) \omega_0(x) - \frac{xy}{BC} \left\{ \omega_2(C) - \omega_0(B) - \omega_1(C) + \omega_0(0) \right\} \\
& + \left\{ \omega_1(y) - \omega_0(0) \right\} + \frac{x}{B} \left\{ \omega_2(y) - \omega_0(B) - \omega_1(y) + \omega_0(0) \right\} \\
& - \frac{y}{C} \left\{ \omega_1(C) - \omega_0(0) - \omega_3(x) \right\}. \qquad (36)
\end{aligned}
$$

From (36) we obtain Laplace equation as follows

$$
\begin{aligned}
\Delta u(x, y) = & \sum_{i=1}^{2M} \sum_{l=1}^{2M} u_{il} \left\{ h_i(x) \left(P_{2l}(y) - \frac{y}{C} P_{2l}(C) \right) + h_l(y) \left(P_{2i}(x) - \frac{x}{B} P_{2i}(B) \right) \right\} \\
& + (1 - \frac{y}{C}) * \omega_0''(x) + \frac{y}{C} * \omega_3''(x) + (1 - \frac{x}{B}) * \omega_1''(y) + \frac{x}{B} * \omega_2''(y). \qquad (37)
\end{aligned}
$$

Replacing u and its Laplacian, given respectively by (36) and (37), in the Eq. (29) leads to

$$
\begin{aligned}
f(x, y) = & \sum_{i=1}^{2M} \sum_{l=1}^{2M} u_{il} \left\{ h_i(x) \left(P_{2l}(y) - \frac{y}{C} P_{2l}(C) \right) + h_l(y) \left(P_{2i}(x) - \frac{x}{B} P_{2i}(B) \right) \right\} \\
& + (1 - \frac{y}{C}) * \omega_0''(x) + \frac{y}{C} * \omega_3''(x) + (1 - \frac{x}{B}) * \omega_1''(y) + \frac{x}{B} * \omega_2''(y) \\
& + k^2 \sum_{i=1}^{2M} \sum_{l=1}^{2M} u_{il} \left\{ \left(P_{2i}(x) - \frac{x}{B} P_{2i}(B) \right) \left(P_{2l}(y) - \frac{y}{C} P_{2l}(C) \right) \right\} \\
& + k^2 (1 - \frac{y}{C}) \omega_0(x) - k^2 \frac{xy}{BC} \left\{ \omega_2(C) - \omega_0(B) - \omega_1(C) + \omega_0(0) \right\} \\
& + k^2 \left\{ \omega_1(y) - \omega_0(0) \right\} + k^2 \frac{x}{B} \left\{ \omega_2(y) - \omega_0(B) - \omega_1(y) + \omega_0(0) \right\} \\
& - k^2 \frac{y}{C} \left\{ \omega_1(C) - \omega_0(0) - \omega_3(x) \right\}. \qquad (38)
\end{aligned}
$$

Express Eq. (38) at a collocation point (x_r, y_s) for $r, s \in \{1, 2, \ldots, 2M\}$, we obtain

$$
\sum_{i=1}^{2M} \sum_{l=1}^{2M} u_{il} D_{ilrs} = T(r, s), \qquad (39)
$$

where

$$
\begin{aligned}
D_{ilrs} = {} & h_i(x)\left(P_{2l}(y) - \frac{y}{C}P_{2l}(C)\right) + h_l(y)\left(P_{2i}(x) - \frac{x}{B}P_{2i}(B)\right) \\
& + k^2\left(P_{2i}(x) - \frac{x}{B}P_{2i}(B)\right)\left(P_{2l}(y) - \frac{y}{C}P_{2l}(C)\right),
\end{aligned}
\tag{40}
$$

and

$$
\begin{aligned}
T(r,s) = {} & f(x_r, y_s) - (1 - \frac{y}{C}) * \omega_0''(x) - \frac{y}{C} * \omega_3''(x) - (1 - \frac{x}{B}) * \omega_1''(y) \\
& - \frac{x}{B} * \omega_2''(y) + k^2 \frac{xy}{BC}\{\omega_2(C) - \omega_0(B) - \omega_1(C) + \omega_0(0)\} \\
& - k^2(1 - \frac{y}{C})\omega_0(x) - k^2\frac{x}{B}\{\omega_2(y) - \omega_0(B) - \omega_1(y) + \omega_0(0)\} \\
& + k^2\frac{y}{C}\{\omega_1(C) - \omega_0(0) - \omega_3(x)\} - k^2\{\omega_1(y) - \omega_0(0)\}.
\end{aligned}
\tag{41}
$$

Equation (39) gives rise to a system of linear equations of order $(2M)^2$.

5 Solving the Linear System

5.1 Regularization and Preconditioning

The fourth-order matrix systems (26) and (39) are transformed into the following form, where only second-order matrices appear

$$
Ac = b, \tag{42}
$$

from which the wavelet coefficients u_{il} reordered in the vector c of order $n = (2M)^2$ can be calculated. The n dimensional vector b is computed from (28) or (41), and the matrix A of order n is given from (27) or (40). In order to circumvent the problems linked to possible ill-conditioning of the matrix A, one can instead of the linear system (42) solve the following regularized problem:

$$
\left[A^T A + \mu I\right] c_\mu = A^T b, \tag{43}
$$

when $\mu = 0$ this system is equivalent to the following system

$$
Dc = b_1, \tag{44}
$$

where $b_1 = A^T b$ and $D = A^T A$.

If this is necessary, i.e. in the case where the regularization does not make it possible to obtain a good approximation or the resolution of the regularized system is very slow, we proceed by preconditioning the system (42) in the following way:

let us define two operators $\mathcal{P}_\gamma$ and $\mathcal{Q}_\delta$ defined respectively for a given matrix A by

$$\mathcal{P}_\gamma(A) = diag\,\{p_1, \ldots, p_n\} \tag{45}$$

and

$$\mathcal{Q}_\delta(A) = diag\,\{q_1, \ldots, q_n\} \tag{46}$$

where p_k and q_k defined by the following formulas:

$$p_k = \gamma \left(\frac{\sum_{i=1}^{n} A_{i1}^2}{\sum_{i=1}^{n} A_{ik}^2} \right)^{(1/2)}, \qquad k = 1, \ldots, n, \tag{47}$$

$$q_k = \delta \left(\frac{\sum_{j=1}^{n} A_{1j}^2}{\sum_{j=1}^{n} A_{kj}^2} \right)^{(1/2)}, \qquad k = 1, \ldots, n. \tag{48}$$

Where γ and δ are amplification factors used to more reduce the condition number. When $\gamma = 1 = \delta$, the resultant matrices QA and AP have the same norm of each row and column respectively.

Now, construct three sequences P_k, Q_k and b_k by

$$P_1 = \mathcal{P}_\gamma(A), \quad A_1 = AP_1 \tag{49}$$

$$Q_k = \mathcal{Q}_\delta(A_{2k-1}),\ k = 1, \ldots \quad P_k = \mathcal{P}_\gamma(A_{2k-2}),\ k = 2, \ldots \tag{50}$$

where

$$A_{2k-1} = A_{2k-2}P_k,\ k = 2, \ldots, M \quad \text{and} \quad A_{2k} = Q_k A_{2k-1}, \quad k = 1, \ldots \tag{51}$$

$$b_0 = b \quad \text{and} \quad b_k = Q_k b_{k-1}, \quad k = 1, \ldots \tag{52}$$

We then call the two-sided k-multi-preconditioned system (TSkMP) of system (26) the following system

$$A_{2k} z_k = b_k,, \quad k = 1, \ldots \tag{53}$$

Once this k-multi-preconditioned system is solved, the vector c solution of the system (26) is obtained by

$$c = P\, z_k \tag{54}$$

Table 1 Results for Example 1 with exact given data and $k^2 = 15$

	γ	MP	Error	Iteration	μ
No.Reg. and No.Prec.	–	–	0.009922198	167	–
Reg. and No.Prec.	–	–	0.009696776	45	1.00E-01
TSkMP	0.4	2	0.006154481	5	1.00E-02

where $P = \prod_{l=1}^{l=k} P_l$. Note that other types of preconditioning such as those cited in [63] and the references therein can be used. The preconditioning developed here is easy to use and very effective as shown by the numerical results presented below. In the next section, we present some numerical results obtained by applying the Haar wavelet approximation method on somme examples.

6 Numerical Computations

Example (1): In this first example, the given data are computed from the exact solution is $u(x, y) = sinh(y) * sin(\pi * x)$. In the Haar wavelets approximation, the maximal level of resolution is taken to be $J = 3$. The tolerance in the stopping criterion in the linear systems solver is $tol = 10^{-6}$. The wavenumber k in the Helmholtz equation (1) and (29) takes values in $\{\sqrt{15}, \sqrt{52}, 10\}$.

In all results we use the following abbreviations: **Reg** to say regularization, **Prec** will designate precondition, **NP** will signify the number of product in the preconditioner and **TSkMP** will designate the two-sided k-multi-preconditioned system.

We observe from Table 1 and Fig. 2 that, the results obtained from the exact given data, when $k^2 = 15$, are very satisfactory even without regularization or preconditioning. The impact of regularization and preconditioning on precision is minimal but the number of iterations is considerably reduced (from 167 to 5).

In the Table 2, we present the results with $g(x)$ computed as the normal derivative of the solution of the direct problem with the data calculated from the solution of example 1 and $k^2 = 15$. We observe that without regularization the method does not produce any acceptable solution. The application of the regularization makes it possible to produce a solution with an error of the order of 10^{-2}.

The application of the preconditioning improves the precision, we find a solution with the same order of precision (error $6 * 10^{-3}$) as for the first case. The number of iteration is divided by 40 but it remains higher than the case with exact data.

Again, when $k^2 = 52$ with exact given data, we get a good solution as can be seen in Fig. 3.

When g is calculated from the direct problem with $k^2 = 52$, it can be seen from Table 3 that the results without regularization without preconditioning are not satisfactory (order error of 23%). But as in the previous case, they are clearly improved

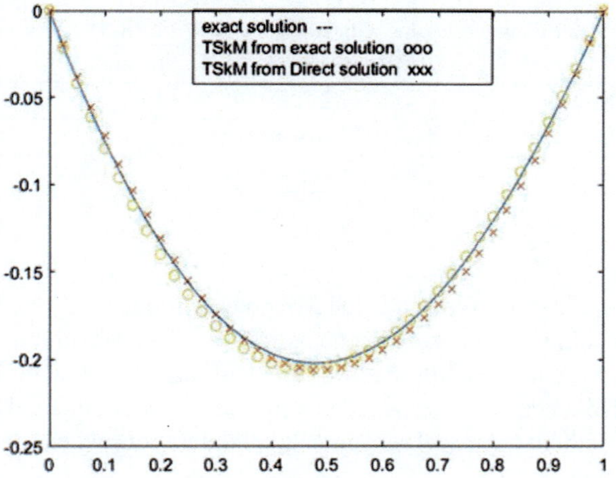

Fig. 2 Exact and computed solutions on Γ_3 for Example 1, with $k^2 = 15$

Table 2 Results for Example 1 using computed given data with $k^2 = 15$

	γ	MP	Error	Iteration	μ
No.Reg. and No.Prec.	–	–	1.898219653	1869	–
Reg. and No.Prec.	–	–	0.012202328	266	1.00E-04
TSkMP	0.1	3	0.006800966	48	1.00E-15

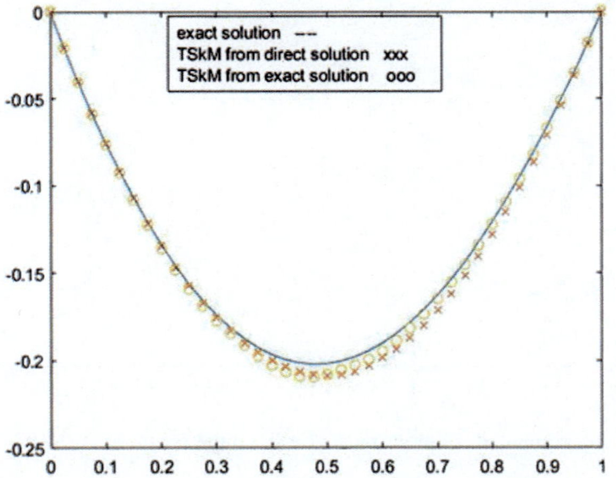

Fig. 3 Exact and computed solutions on Γ_3 for Example 1 with $k^2 = 52$

Table 3 Results for Example 1 using computed given data on Γ_3 with $k^2 = 52$

	γ	MP	Error	Iteration	μ
No.Reg. and No.Prec.	–	–	0.231224892	1065	–
Reg. and No.Prec.	–	–	0.019788832	250	1.00E-04
TSkMP	0.1	1	0.008443771	3	1.00E-02

with the regularization. When the TSkMP preconditioner is applied we reach a good approximed solution, the error is equals to 8×10^{-3}. We find a solution with the same order of precision as for the first case. But it remains higher than the case with exact data. This accuracy is obtained in only 3 iterations, compared to 1065 iterations for the case without any regularization and 300 iterations when regularizing without preconditioning.

For the case where the wavenumber is given by $k^2 = 100$, the same conclusions as for the previous cases can be drawn by observing Fig. 4 and the Table 4.

Example (2): In this example,the given data are computed from the exact solution $u(x, y) = sinh(p * y) * sin(p * pi * x)/p^2$. The other data are kept as in Example 1.

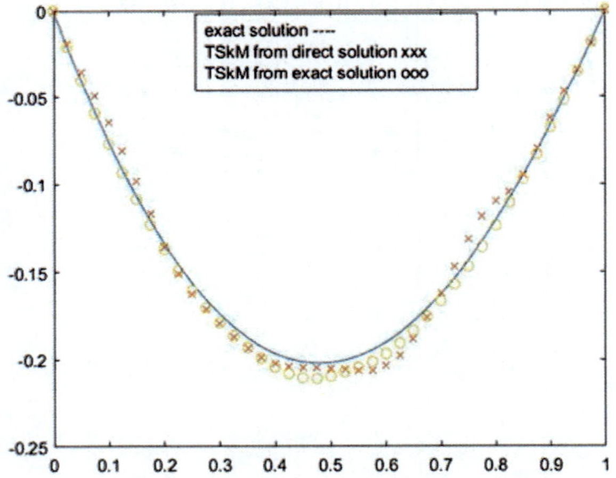

Fig. 4 Exact and computed solutions on Γ_3 for Example 1 with $k^2 = 100$

Table 4 Results for Example 1 using computed given data on Γ_3 with $k^2 = 100$

	γ	MP	Error	Iteration	μ
No.Reg. and No.Prec.	–	–	0.038537796	790	–
Reg. and No.Prec.	–	–	0.034444278	34	1.00E-04
TSkM	0.4	5	0.014321817	9	1.00E-08

Table 5 Results for Example 2 using computed given data on Γ_3 with $k^2 = 15$

	γ	MP	Error	Iteration	μ
No.Reg. and No.Prec.	–	–	8.179606174	1873	–
Reg. and No.Prec.	–	–	0.108911018	247	1.00E-03
TSkM	0.1	2	0.093508507	27	1.00E-10

In all the numerical experiments that we have developed, the results for the different wave numbers follow the same order as in example 1. We therefore choose to present only the numerical results corresponding to the case $k^2 = 15$ for Example 2 and Example 3 below.

For precision and convergence we only present the results for g calculated from the direct problem. The normal derivative being calculated and not exact allows us to simulate real situations where the data is measured and therefore contains noise compared to exact data. This allows us to test the stability of the approximation produced

Concerning Example 2, the method without regularization without preconditioning does not produce a good solution, see Table 5. The regularization without preconditioning makes it possible to reach a precision of the order of 11% and reduces the number of iterations from 1873 to 247. By introducing an additional preconditioning, we obtain a fairly precise approximation error equal to 9% in 27 iterations.

We observe from Fig. 5 that for this example also the approximation obtained with exact data always remains very satisfactory even without regularization or preconditioning. When the data are disturbed, the preconditioning makes it possible to obtain a stable solution in very few iterations.

Example (3): The data for this example was calculated from the function $u(x, y) = sin(x) * sin(y) * cos((pi/2) * y) * cos((pi/2) * x)$. The other parameters remain the same as in the first two examples.

Again, the results presented in Table 6 and Fig. 6 show that the solution obtained with exact data is always of good quality. When the data is noisy, the regularization of the linear system combined with the preconditioning makes it possible to obtain a stable solution.

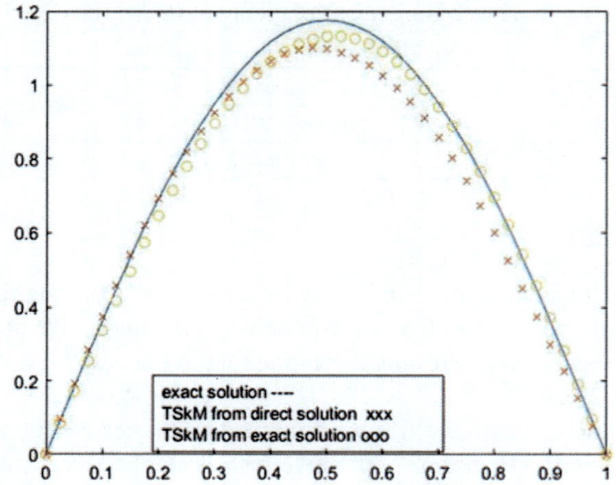

Fig. 5 Exact and computed solutions on Γ_3 for Example 2 with $k^2 = 15$

Table 6 Results for Example 3 using computed given data on Γ_3 with $k^2 = 15$

	γ	MP	Error	Iteration	μ
No.Reg. and No.Prec.	–	–	0.413675761	781	–
Reg. and No.Prec.	–	–	0.08351874	128	1.00E-03
TSkM	0.1	3	0.076871767	50	1.00E-15

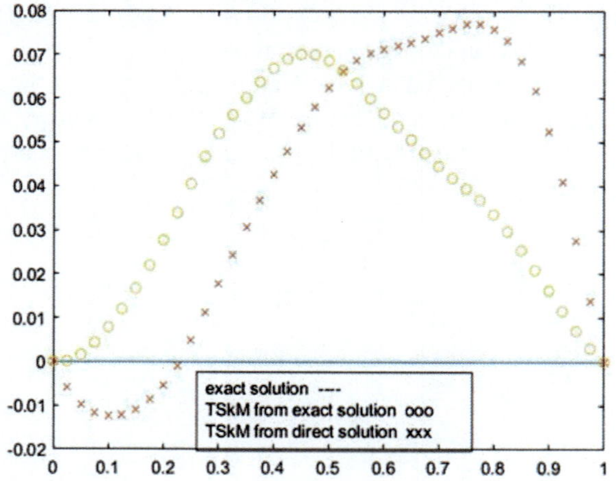

Fig. 6 Exact and computed solutions on Γ_3 for Example 3 with $k^2 = 15$

7 Conclusion

We have presented a method for solving inverse Cauchy problems governed by the Helmholtz equation using Haar wavelets to reconstruct the unknown boundary condition on an accessible boundary from the given data on another part of the boundary.

The Haar wavelet method has several advantages over other numerical methods for solving the Cauchy problem. First, it is a direct semi-analytical method which is computationally efficient. A Second advantage is the ease of implementation, as the method does not require the use of complex numerical algorithms. Finally, the method is easily adaptable to problems in higher dimensions, making it a powerful tool for solving a wide range of inverse problems. However, the Haar wavelet method also has some limitations. For example, the method is not always stable and can lead to numerical instability when applied to certain types of Cauchy boundary conditions. Despite these limitations which can be circumvented using the proposed preconditioning, the Haar wavelet method remains a valuable tool for solving a wide range of inverse problems. With continued research and development, it is likely that the method will become even more powerful and widely used in the future.

Overall, the method based on using Haar wavelets to solve inverse Cauchy problems is a powerful technique that can be used to approximate solutions to a wide range of inverse problems governed by different class of partial differential equations. By using the proposed preconditioner, it is possible to obtain accurate and stable solutions even for ill-conditioned systems.

Acknowledgements This work was developed during the visit of Sudad Musa Rashid, at the **Laboratoire de Mathématiques Jean Leray, Nantes Université (France), CNRS UMR 6629**. This scientific stay was partially funded by **Fédération de Recherche Mathématiques des Pays de Loire-FR CNRS 2962**.

References

1. C.H. Huang, W.C. Chen, A three-dimensional inverse forced convection problem in estimating surface heat flux by conjugate gradient method. Int. J. Heat Mass Transf. **43**(17), 3171–3181 (2000)
2. A. Nachaoui, An improved implementation of an iterative method in boundary identification problems. Numer. Algorithms **33**(1–4), 381–398 (2003)
3. T. Regińska, K. Regiński, Approximate solution of a Cauchy problem for the Helmholtz equation. Inverse Probl. **22**(3), 975–989 (2006)
4. A. Arsenashvili, A. Nachaoui, T. Tadumadze, On approximate solution of an inverse problem for linear delay differential equations. Bull. Georg. Natl. Acad. Sci. (N.S.) **2**(2), 24–28 (2008)
5. C.L. Fu, X.L. Feng, Z. Qian, The Fourier regularization for solving the Cauchy problem for the Helmholtz equation. Appl. Numer. Math. **59**(10), 2625–2640 (2009)
6. A. Chakib, A. Nachaoui, A. Zeghal, A shape optimization approach for an inverse heat source problem. Int. J. Nonlinear Sci. **12**(1), 78–84 (2012)
7. A. Boulkhemair, A. Nachaoui, A. Chakib, A shape optimization approach for a class of free boundary problems of Bernoulli type. Appl. Math. **58**(2), 205–221 (2013)

8. M.M. Lavrentiev, *Some Improperly Posed Problems of Mathematical Physics* (Springer Science & Business Media, 2013)
9. A. Chakib, M. Johri, A. Nachaoui, M. Nachaoui, On a numerical approximation of a highly nonlinear parabolic inverse problem in hydrology. Ann. Univ. Craiova Ser. Math. Inf. **42**, 192–201 (2015)
10. V. Isakov, Inverse problems for partial differential equations, in *Applied Mathematical Sciences*, vol. 127 (Springer, Cham, 2017)
11. C.S. Liu, F. Wang, A meshless method for solving the nonlinear inverse Cauchy problem of elliptic type equation in a doubly-connected domain. Comput. Math. Appl. **76**, 1831–1852 (2018)
12. A. Bergam, A. Chakib, A. Nachaoui, M. Nachaoui, Adaptive mesh techniques based on a posteriori error estimates for an inverse Cauchy problem. Appl. Math. Comput. **346**, 865–878 (2019)
13. K.A. Berdawood, A. Nachaoui, R. Saeed, M. Nachaoui, F. Aboud, An alternating procedure with dynamic relaxation for Cauchy problems governed by the modified Helmholtz equation. Adv. Math. Model. Appl. **5**(1), 131–139 (2020)
14. F. Wang, Y. Gu, W. Qu, C. Zhang, Localized boundary knot method and its application to large-scale acoustic problems. Comput. Methods Appl. Mech. Eng. **361**, 112729 (2020)
15. A. Chakib, A. Hadri, A. Laghrib, On a multiscale analysis of an inverse problem of nonlinear transfer law identification in periodic microstructure. Ann. Univ. Craiova Ser. Math. Inf. **51**, 102985 (2020)
16. F. Aboud, A. Nachaoui, M. Nachaoui, On the approximation of a Cauchy problem in a non-homogeneous medium. J. Phys.: Conf. Ser. **1743**(1), 012003 (2021)
17. A. Nachaoui, M. Nachaoui, A. Chakib, M. Hilal, Some novel numerical techniques for an inverse Cauchy problem. J. Comput. Appl. Math. **381**(113030)
18. A. Nachaoui, A. Laghrib, M. Hakim, Mathematical control and numerical applications, in *Springer Proceedings in Mathematics & Statistics*, vol. 372 (Springer, Cham, 2021)
19. M. Nachaoui, A. Nachaoui, T. Tadumadze, On the numerical approximation of some inverse problems governed by nonlinear delay differential equation. RAIRO Oper. Res. **56**, 1553–1569 (2022)
20. L. Afraites, C. Masnaoui, M. Nachaoui, Shape optimization method for an inverse geometric source problem and stability at critical shape. Discret. Contin. Dyn. Syst. Ser. S **15**(1), 1–21 (2022)
21. H. Ouaissa, A. Chakib, A. Nachaoui, M. Nachaoui, On numerical approaches for solving an inverse cauchy stokes problem. Appl. Math. Optim. **85**(1). https://doi.org/10.1007/s00245-022-09833-8
22. A. Ellabib, A. Nachaoui, A. Ousaadane, Convergence study and regularizing property of a modified Robin-Robin method for the Cauchy problem in linear elasticity. Inverse Probl. **38**, 075007 (2022)
23. A. Nachaoui, Cauchy's problem for the modified biharmonic equation: ill-posedness and iterative regularizing methods, in *New Trends of Mathematical Inverse Problems and Applications, Springer Proceedings in Mathematics & Statistics* (Springer, Cham, 2023)
24. A. Nachaoui, F. Aboud, Solving geometric inverse problems with a polynomial based meshless method, in *New Trends of Mathematical Inverse Problems and Applications, Springer Proceedings in Mathematics & Statistics* (Springer, Cham, 2023)
25. A. Nachaoui, M. Nachaoui, T. Tadumadze, Meshless methods to noninvasively calculate neurocortical potentials from potentials measured at the scalp surface, in *New Trends of Mathematical Inverse Problems and Applications, Springer Proceedings in Mathematics & Statistics* (Springer, Cham, 2023)
26. J. Blum, *Numerical Simulation and Optimal Control in Plasma Physics with Applications to Tokamaks* (Wiley, Chichester, 1989)
27. A. Ellabib, A. Nachaoui, An iterative approach to the solution of an inverse problem in linear elasticity. Math. Comput. Simul. **77**, 189–201 (2008)

28. D. Maxwell, M. Truffer, S. Avdonin, M. Stueferv, An iterative scheme for determining glacier velocities and stresses. J. Glaciol. **54**(188), 888–898 (2008)
29. J.C. Liu, T. Wei, A quasi-reversibility regularization method for an inverse heat conduction problem without initial data. Appl. Math. Comput. **219**(23), 10866–10881 (2013)
30. A. Chakib, A. Nachaoui, M. Nachaoui, H. Ouaissa, On a fixed point study of an inverse problem governed by stokes equation. Inverse Probl. **35**, 015008 (2019)
31. E. Hernandez-Montero, A. Fraguela-Collar, J. Henry, An optimal quasi solution for the Cauchy problem for Laplace equation in the framework of inverse ECG. Math. Model. Nat. Phenom. **14**
32. M. Malovichko, N. Koshev, N. Yavich, A. Razorenova, M. Fedorov, Electroencephalographic source reconstruction by the finite-element approximation of the elliptic Cauchy problem. EEE Trans. Biomed. Eng. **68**(6), 1019–1811 (2020)
33. K.A. Berdawood, A. Nachaoui, M. Nachaoui, F. Aboud, An effective relaxed alternating procedure for Cauchy problem connected with Helmholtz equation. Numer. Methods Part. Differ. Equ. 1–27 (2021)
34. A. Ellabib, A. Nachaoui, A. Ousaadane, Mathematical analysis and simulation of fixed point formulation of Cauchy problem in linear elasticity. Math. Comput. Simul. **187**, 231–247 (2021)
35. S.M. Rasheed, A. Nachaoui, M.F. Hama, A.K. Jabbar, Regularized and preconditioned conjugate gradient like-methods methods for polynomial approximation of an inverse Cauchy problem. Adv. Math. Model. Appl. **6**(2), 89–105 (2021)
36. F. Aboud, I.T. Jameel, A.F. Hasan, B.K. Mostafa, A. Nachaoui, Polynomial approximation of an inverse Cauchy problem for Helmholtz type equations. Adv. Math. Model. Appl. **7**(3), 306–322 (2022)
37. A. Nachaoui, Iterative methods for inverse problems subject to the convection-diffusion equation, in *New Trends of Mathematical Inverse Problems and Applications, Springer Proceedings in Mathematics & Statistics* (Springer, Cham, 2023)
38. A. Nachaoui, M. Nachaoui, M.A. Hilal, A new approach for solving an inverse Cauchy problem based on bfgs method, in *New Trends of Mathematical Inverse Problems and Applications, Springer Proceedings in Mathematics & Statistics* (Springer, Cham, 2023)
39. J. Hadamard, *Lectures on Cauchy's Problem in Linear Partial Differential Equations* (Yale University Press, New Haven, 1923)
40. P. Dvalishvili, A. Nachaoui, M. Nachaoui, T. Tadumadze, On the well-posedness of the Cauchy problem for a class of differential equations with distributed delay and the continuous initial condition. Proc. Inst. Math. Mech. Natl. Acad. Sci. Azerbaijan **43**, 146–160 (2017)
41. F. Berntsson, V.A. Kozlov, L. Mpinganzima, B.O. Turesson, Iterative Tikhonov regularization for the Cauchy problem for the Helmholtz equation. Comput. Math. Appl. **73**, 163–172 (2017)
42. A.L. Qian, X.T. Xiong, Y.-J. Wu, On a quasi-reversibility regularization method for a Cauchy problem of the Helmholtz equation. J. Comput. Appl. Math. **233**(8), 1969–1979 (2010)
43. S. Yarmukhamedov, I. Yarmukhamedov, Cauchy problem for the Helmholtz equation, in *Ill-Posed and Non-Classical Problems of Mathematical Physics and Analysis, Inverse Ill-Posed Problem Series*, VSP, Utrecht (2003), pp. 143–172
44. Q. Hua, Y. Gu, W. Qu, W. Chen, C. Zhang, A meshless generalized finite difference method for inverse Cauchy problems associated with three-dimensional inhomogeneous Helmholtz-type equations. Eng. Anal. Bound. Elem. **82**, 162–171 (2017)
45. Z. Qian, X. Feng, A fractional Tikhonov method for solving a Cauchy problem of Helmholtz equation. Appl. Anal. **96**, 1656–1668 (2017)
46. F. Yang, P. Zhang, X.X. Li, The truncation method for the Cauchy problem of the inhomogeneous Helmholtz equation. Appl. Anal. **98**, 991–1004 (2019)
47. F. Wang, Z. Chen, P.-W. Li, C.-M. Fan, Localized singular boundary method for solving Laplace and Helmholtz equations in arbitrary 2d domains. Eng. Anal. Bound. Elem. **129**, 82–92 (2021)
48. K. Berdawood, A. Nachaoui, R. Saeed, M. Nachaoui, F. Aboud, An efficient DN alternating algorithm for solving an inverse problem for Helmholtz equation. Discret. Contin. Dyn. Syst. Ser. S

49. K. Berdawood, A. Nachaoui, M. Nachaoui, An accelerated alternating iterative algorithm for data completion problems connected with Helmholtz equation. Stat., Optim. Inf. Comput. **11**(1), 2–21 (2023). https://doi.org/10.1002/num.22793
50. C.F. Chen, C.H. Hsiao, Haar wavelet method for solving lumped and distributed parameter systems. Proc. Control Theory Appl. **144**(1), 87–94 (1997)
51. C.H. Hsiao, W.J. Wang, Haar wavelet approach to nonlinear stiff systems. Math. Comput. Simul. **57**, 347–353 (2001)
52. R.S. Stanković, B.J. Falkowski, The Haar wavelet transform: its status and achievements. Comput. Electr. Eng. **29**, 24–44 (2003)
53. U. Lepik, Numerical solution of differential equations using Haar wavelets. Math. Comput. Simul. **68**(2), 127–143 (2005)
54. S. Yousefi, Legendre wavelets method for solving differential equations of lane-Emden type. Appl. Math. Comput. **181**(2), 1417–1422 (2006)
55. U. Lepik, Haar wavelet method for solving stiff differential equations. Math. Model. Anal. **14**, 467–481 (2009)
56. R. Dai, J.E. Cochran, Wavelet collocation method for optimal control problems. J. Optim. Theory Appl. **143**(2), 265–278 (2009)
57. G. Hariharan, K. Kannan, K.R. Sharma, Haar wavelet method for solving fisher's equation. Appl. Math. Comput. **211**(2), 284–292 (2009)
58. R. Kalpana, S. Raja Balachandar, Haar wavelet method for the analysis of transistor circuits. Int. J. Electron. Commun. (AEU) **61**, 589–594 (2007)
59. U. Lepik, Numerical solution of evolution equations by the Haar wavelet method. Appl. Math. Comput. **185**(1), 695–704 (2007)
60. A. Nachaoui, E.S. Al-Rawi, A.F. Qasim, Solving three dimensional and time depending PDEs by Haar wavelets method. Open Access Libr. J. **5**(5), 1–18 (2018)
61. S.M. Rashid, A. Nachaoui, A Haar wavelets-based direct reconstruction method for the Cauchy problem of the Poisson equation. Discret. Contin. Dyn. Syst. Ser. S
62. A. Haar, Zur theorie der orthogonalen funktion systeme. Math. Annal. **69**, 331–371 (1910)
63. K. Boumzough, A. Azzouzi, A. Bouidi, The incomplete LU preconditioner using both CSR and CSC formats. Adv. Math. Model. Appl. **7**(2), 156–167 (2022)

Meshless Methods to Noninvasively Calculate Neurocortical Potentials from Potentials Measured at the Scalp Surface

Abdeljalil Nachaoui, Mourad Nachaoui, and Tamaz Tadumadze

Abstract Noninvasive measurement of neurocortical potentials using electroencephalography (EEG) is a valuable tool in neuroscience research and clinical practice. However, accurate estimation of neurocortical potentials from scalp potentials is a challenging problem, due to the complex and ill-posed nature of the forward and inverse problems. In this paper, we present an approach based on the annular model and polynomial expansion method to solve the Cauchy problem associated with noninvasive calculation of neurocortical potentials. We derive the mathematical formulas and discuss the numerical implementation of the method, and demonstrate its effectiveness using numerical simulations.

Keywords Data reconstruction · Inverse problem · Cauchy problem · Meshless methods · Neurocortical potentials · Polynomial expansion method

1 Introduction

The human brain generates electrical and magnetic fields as a result of the activity of neurons. These fields can be measured noninvasively using EEG or MEG, which provide a convenient and safe way to study brain function and dysfunction. However, the signals measured at the scalp surface are a mixture of the underlying neurocortical potentials and other sources, such as noise, artifacts, and volume

A. Nachaoui (✉)
Laboratoire de Mathématiques Jean Leray, Nantes Université, Nantes, France
e-mail: Abdeljalil.Nachaoui@univ-nantes.fr

M. Nachaoui
(EMI), Université Sultan Moulay Slimane, Béni-Mellal, Morocco
e-mail: m.nachaoui@usms.ma

T. Tadumadze
I. Vekua Institute of Applied Mathematics, I. Javakhishvili Tbilisi State University, Tbilisi, Georgia
e-mail: tamaz.tadumadze@tsu.ge

A. Laghrib et al. (eds.), *New Trends of Mathematical Inverse Problems and Applications*, Springer Proceedings in Mathematics & Statistics 428,
https://doi.org/10.1007/978-3-031-33069-8_7

conduction effects. Therefore, accurate estimation of neurocortical potentials from scalp potentials requires solving the forward problem, which relates the sources to the measurements, and the inverse problem, which reconstructs the sources from the measurements.

Electroencephalography (EEG), Magnetoencephalography (MEG) are medical imaging techniques, which measure the variation of physical and/or physiological parameters related to the perception or the realization of a cognitive task (linguistic tasks, memory, face recognition...) or in relation to a local cerebral dysfunction. EEG and MEG simultaneously measure electric and magnetic fields at EEG are therefore measurements on the surface of the head indicating the electrical activity of current generators. To perform these measurements, it is necessary to model the head as a conductive volume and to apply the laws of electromagnetism to estimate the volume distributions of the current sources. The problem that arises in this area is to locate the sources that caused the EEG and MEG signals and this to know and where the stages of information processing take place in the brain [1], In other words fact that the EEG and the MEG have an excellent temporal resolution for the observation of electrophysiological phenomena, the question which then arises is whether it is possible to estimate the spatial and temporal localization of the different neurons in activation which have participate in activation spots in the brain.

The two fundamental problems that arise in electrophysiology can be defined as follows:

1. The direct problem: given a source q, find the potential ϕ for all r in $\Omega \subset \mathbb{R}^3$. The solution is unique and determined by the boundary conditions.
2. The inverse problem: the potential ϕ being given in $\Omega \subset \mathbb{R}^2$ (closed surface containing sources q), find q with $q = 0$ on the surface and outside the surface. This problem cannot be solved by conventional methods.

The electric potential associated with the brain sources fulfills the Laplace's equation there since the exterior compartments (tissues) lack any sources giving the objective signal. In order to map the data from the scalp to the brain surface, it is possible to state the Cauchy problem for the Laplace's equation by measuring the potential on the portion of the head surface. Due to the higher quality and spatial resolution of the inputted data, this mapping is an auxiliary issue that enables a sufficient improvement in source localization accuracy.

The Cauchy problem for the data mapping in the context of EEG was considered in [2], using a minimization procedure with the boundary elements method (BEM) approximation and in [3] where authors used a technique based on the mixed quasi-reversibility (MQR) method for linear finite elements.

With the objective of detecting the active parts of the cortex, we are interested in the simulation and numerical aspects of solving this inverse Cauchy problem as an intermediate procedure to propagate the data from the scalp to the cortex before employing other processing techniques. To this end, we proposed to apply successive method used for solving Cauchy problems in non-homogeneous bounded domain [4] with a polynomyal expansion [5, 6].

The proposed method can serve as the basis for fast and accurate algorithms for the analysis of real EEG data.

The paper is organized as follows. In Sect. 2 we briefly recall the head models used in this field as well as the notion of inverse problems and the resolution techniques. Section 3, is devoted to the conductivity model of the human head determining the EEG as well as the presentation of the equations for the sourceless compartments of the head . We describe in Sect. 4 the successive approach as well as the polynomial expansion used to solve the Cauchy problem. We end this section by presenting some important notes about our implementation. Section 5 presents some numerical results. Concluding remarks are given in Sect. 6.

2 Overview of the Problem

The noninvasive calculation of neurocortical potentials from potentials measured at the scalp surface is an important problem in neuroscience and clinical applications. Neurocortical potentials, also known as electroencephalography (EEG) or magnetoencephalography (MEG) signals, provide valuable information about brain activity and function. Noninvasive techniques for measuring these signals are preferred over invasive methods, which require the insertion of electrodes or sensors directly into the brain tissue.

2.1 Modeling the Geometry of the Head

Two models are used to represent the head, the spherical model and the realistic model. The first model is the oldest and the most used. The second model has mainly made its appearance with the advancement of measurement techniques such as MRI and it is proposed according to two volumetric and surface approaches. We content ourselves with describing the first model that we are going to use in this study.

The Spherical Model This is the first model to be used and the most simplistic. It considers the head as a series of concentric spheres with homogeneous and isotropic conductivities. The most simplified models consist of three layers; the scalp the skull and the intracranial volume (or just the brain). The 4-layer models add a layer for cerebrospinal fluid between the bone and the brain. The most complete models include five layers, they add gray matter and white matter. The distance between the different spheres corresponds to the average real distance between the tissues represented. For this model, we know an analytical solution for the direct problem. Thus, the spherical model is often used as a reference for a first validation and evaluation of numerical methods. The three-layer model, studied for the first time in [7], is the most used in clinics and research. In this model, three concentric spheres are used with the following typical radius values: $R_{brain} = 8cm$, $R_{skull} = 8.5cm$ and

$R_{scalp} = 9.2cm$ (or also after scaling: $(0.87, 0.92$ and $1)$ used in an attempt to better represent the geometry of the tissues of the head [8].

2.2 *Inverse Problems*

Inverse problems are of interest to many fields in everyday life, they are of interest to thermicians, mathematicians, statisticians, geophysicists, mechanics, engineers, electronics engineers [9–33], and essentially the field of medical imaging to which we are concerned [2, 3, 21, 34–38].

The purpose of the inverse methods is to determine quantities that are difficult to measure from easily observable quantities (here, the measurements of potentials at the surface of the scalp).

In a general way, one can say that an inverse problem is reduced to the extraction of the maximum of information contained in observations of such a physical system. So the inverse problems in EEG are considered as a better solution to the brain dynamics estimation problem Indeed, the purpose of the inverse problem in the EEG is to find the neuron bundles. In the origin of the electrical potentials measured on the surface of the head. It transforms a mixture of brain electrical activity obtained on the surface of the scalp into a local estimate of each significant contribution. So it is a localization of the different current sources that causes disturbances inside the brain such as pilepsy for example.

We propose a method based on the annular model and polynomial expansion to noninvasively calculate neurocortical potentials from scalp potentials. The annular model represents the head as an annulus with two concentric spheres that approximate the brain and skull, respectively. The polynomial expansion approximates the potential and its normal derivative as a sum of two-dimensional polynomials in the Cartesian coordinates of the annular model. By expanding the potential and its normal derivative in this way, we can express the Cauchy data corresponding to the scalp potentials and their normal derivatives in terms of the expansion coefficients. Solving the resulting linear system of equations for the expansion coefficients yields an approximation of the neurocortical potentials in the annular model.

3 Theoretical Background

The noninvasive calculation of neurocortical potentials from scalp potentials is governed by the Laplace equation on a head model. Let $\Omega = \cup_{j=0}^{2} \Omega_j$ denote the head model, which we assume is a 3-layer head model (see Fig. 1), representing from inside to outside: Ω_0 for the Brain, Ω_1 for the Skull and Ω_2 for the Scalp, each of which is considered to be homogeneous and isotropic (assuming constant scalar conductivity). The Laplace equation is given by:

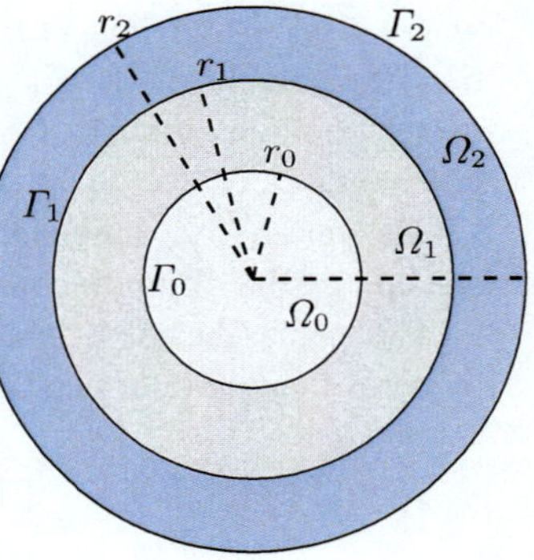

Fig. 1 Two-annulus head model with radii r_0, r_1, and r_2 for the Brain, Skull, and Scalp layers, respectively

$$\nabla \cdot (\sigma \nabla u(x, y)) = 0 \quad \text{in } \Omega, \tag{1}$$

where $\nabla = (\frac{\partial}{\partial x}, \frac{\partial}{\partial y})^T$ is the gradient operator and where $u(x, y)$ is the potential at the point (x, y) in the head model.

Considering the quasi-static approximation [34, 37, 39], the conductivity $\sigma(x, y), (x, y) \in \Omega$ is assumed to be a piecewise-constant distribution over the volume under consideration (head). The conductivity of the media outside the head expected to be zero:

$$\sigma(x, y) = \begin{cases} \sigma_i, & (x, y) \in \Omega_i \subset \Omega; \\ 0, & (x, y) \notin \Omega. \end{cases} \tag{2}$$

3.1 The Associated Cauchy Problem

We denote by Γ_j the interface between Ω_j, Ω_{j+1} for $j = 0, 1$ and by Γ_2 the external surface of the Scalp. We denote by S the part of Γ_2 on which the measurements of the electric potential are obtained by the EEG.

To obtain the neurocortical potentials from the scalp potentials, we need to solve the Cauchy problem associated with this problem. The Cauchy problem requires that we know the potential and the normal derivative of the potential on S the part of Γ_2 on which the measurements of the electric potential are obtained by the EEG. Mathematically, the Cauchy problem is defined as:

$$\nabla^2 u(x, y) = 0 \quad \text{in } \Omega, \tag{3}$$

$$u(x, y) = f(x, y) \quad \text{on } S \tag{4}$$

$$\partial_\nu u(x, y) = g(x, y) \quad \text{on } S \tag{5}$$

where $f(x, y)$ and $g(x, y)$ are the given potential and normal derivative of the potential, respectively, on the S a part of scalp surface Γ_2. Here, ν denotes the unit normal vector pointing outwards from the boundary and $\partial_\nu u$ is the normal derivative of the potential.

The variability of the electrical properties of the tissues of the head is taken into account by the different conductivities σ_i in each of the tissues.

The media (the head) being heterogeneous, the problem (3)–(5) must be completed by the following interface conditions:

$$\begin{cases} [u] & = & 0 & on & \Gamma_1 \\ [\sigma \partial_\nu u] & = & 0 & on & \Gamma_1, \end{cases} \tag{6}$$

where $\sigma = \sigma_i$ in Ω_i.

We therefore consider the problem of determining the electric potential and flux on the surface Γ_0 of the brain from the Cauchy data available on the surface of the scalp Γ_2 described by the following Cauchy problem:

$$\begin{cases} -\Delta u & = & 0 & dans & \Omega_i, i = 2, 1 \\ u & = & f & sur & S \subset \Gamma_2 \\ \partial_\nu u & = & g & sur & \Gamma_2 \\ [u] & = & 0 & sur & \Gamma_1 \\ [\sigma \partial_\nu u] & = & 0 & sur & \Gamma_1 \end{cases} \tag{7}$$

This is an inverse Cauchy problem in an inhomogeneous domain formed by two consecutive annulus. To solve this problem, we will initially use a method of successive resolution of two Cauchy problems, proposed in [4] and the mesheless method based on a polynomial expansion used in [5, 6].

The class of inverse problems called Cauchy has attracted considerable attention from researchers and engineers due to its wide applications in fields ranging from physics and engineering to medicine and the environment [31, 32, 38, 40–45]. It is well known that Cauchy problems are ill-posed [46–48]. This is why any resolution method must take this character into account. In [49, 50] authors used a relaxed alternating iterative methode which reduces the resolution of the problem to solving alternatively two well-posed problems. The regularizing character as well as convergence acceleration intervals have been determined in [51] for the Poisson equation and in [33] for the helmholtz equation. This technique was adopted in ([52]) and ([53]) for a nonlinear elliptic problem, then it was applied for a large class of inverse boundary problems such as structural mechanics [54–56], electromagnetism [57, 58], fluid mechanics [59], inhomogeneous medium [4] and the convection-diffusion equation [60].

Other methods have been developed to solve this kind of problem, the reader can consult for example [61–69] and the references therein.

4 Successive Approach

This approach consists of solving the initial problem (7) in two steps. We first solve a Cauchy problem in the domain formed by the scalp bounded by Γ_2 and Γ_1. We thus transfer the Cauchy data from the boundary Γ_2 to the boundary Γ_1. Then, we solve a second Cauchy problem in the skull, i.e. the domain bounded by Γ_1 and the surface of the brain Γ_0 from the data collected on the scalp-cranium interface Γ_1. The two problems are written:

$$(PC_{Scalp}) \begin{cases} -\Delta u_2 = 0 & \text{in } \Omega_2 \\ u_2 = f & \text{on } S \subset \Gamma_2 \\ \partial_\nu u_2 = g & \text{sur } \Gamma_2 \end{cases} \tag{8}$$

Solving the problem (PC_{Scalp}) allows us to obtain the traces of the potential and the flux on Γ_1: $u_{2|\Gamma_1}$ and $\partial_\nu u_{2|\Gamma_1}$

$$(PC_{Skull}) \begin{cases} -\Delta u_1 = 0 & \text{in } \Omega_1 \\ u_1 = u_2 & \text{on } \Gamma_1 \\ \partial_\nu u_1 = -\frac{\sigma_2}{\sigma_1} \partial_\nu u_2 & \text{on } \Gamma_1 \end{cases} \tag{9}$$

Solving the problem (PC_{Skull}) allows us to obtain the traces of the potential and the flux on the surface of the brain Γ_0. This makes it possible to determine the distribution of the potential in the brain by solving another Cauchy problem in Ω_0 since we obtain two conditions on Γ_0 the boundary of Ω_0.

In both cases, we must solve a Cauchy problem in a annulus of the following form:

$$(PC_{suc}) \begin{cases} -\Delta u = 0 & \text{in } \Omega \\ u = f & \text{sur } S \subset \Gamma_e, \\ \partial_\nu u = g & \text{sur } \Gamma_e, \end{cases} \tag{10}$$

where Ω is an annulus in $\mathbb{R}^2$ representing the domain between two layers of the head, Γ_i the inner surface and Γ_e the outer surface of the layer. S is part of Γ_e for the first annulus (scalp) and $S = \Gamma_e$ for the second annulus (skull). f and g are the Cauchy data. They are supposed to be compatible, ie being respectively the trace and the normal trace of a harmonic function on S and Γ_e.

We can use the polynomial expansion of the potential to solve the Cauchy problem associated with noninvasive calculation of neurocortical potentials from scalp potentials, as described in the previous sections.

4.1 Polynomial Expansion Method

To solve the Cauchy problem associated with noninvasive calculation of neurocortical potentials from scalp potentials, we use a polynomial expansion method. Specifically, we expand the potential $u_2(x, y)$ in terms of a polynomial series:

$$u(x, y) = \sum_{i=1}^{m} \sum_{j=1}^{i} c_{ij} x^{i-j} y^{j-1} \tag{11}$$

where m is the order of the polynomial expansion and c_{ij} are the unknowns expansion coefficients that need to be determined.

Substituting the polynomial expansions (11) into the Laplace equation and the boundary conditions of problem (10) and using some collocation points, we obtain a system of linear equations for the expansion coefficients c_{ij}. The details of this derivation are given in the folowing section.

This method has the advantage of reducing the Cauchy problem to a finite-dimensional problem, which can be solved numerically. Moreover, it provides a systematic way of choosing the number of terms m in the polynomial expansion to achieve a desired accuracy. However, the method may suffer from numerical instability when the number of terms is large, but this can be circumvented using the technique described in Sect. 4.3.

4.2 Substitution of the Polynomial Expansions into the Laplace Equation and Boundary Conditions

In order to approach the solution of the problem (9), one must first approach the solution of (8) to obtain the boundary conditions u_2 and $\partial_\nu u_2$ on Γ_1.

First, we do a polynomial expansion of u_2:

$$u_2(x, y) = \sum_{i=1}^{m} \sum_{j=1}^{i} c_{ij}^{(2)} x^{i-j} y^{j-1} \tag{12}$$

We calculate the derivative of u with respect to x and y, we obtain:

$$\partial_x u_2(x, y) = \sum_{i=1}^{m} \sum_{j=1}^{i} c_{ij}^{(2)} (i - j) x^{i-j-1} y^{j-1} \tag{13}$$

$$\partial_y u_2(x, y) = \sum_{i=1}^{m} \sum_{j=1}^{i} c_{ij}^{(2)} (j - 1) x^{i-j} y^{j-2} \tag{14}$$

Thus, we expand the normal derivative of the potential $\partial_\nu u_2(x, y)$ as:

$$\partial_\nu u_2(x, y) = \cos(\theta) \sum_{i=1}^{m} \sum_{j=1}^{i} c_{ij}^{(2)} (i - j) x^{i-j-1} y^{j-1} + \sin(\theta) \sum_{i=1}^{m} \sum_{j=1}^{i} c_{ij}^{(2)} (j - 1) x^{i-j} y^{j-2} \quad (15)$$

where θ is the angle between the outward normal vector at a point on the boundary and the positive x-axis.

Similarly, we expand the Laplacian of $u_2(x, y)$ as

$$\Delta u_2(x, y) = \sum_{i=1}^{m} \sum_{j=1}^{i} \left((i - j)(i - j - 1) x^{i-j-2} y^{j-1} + (j - 1)(j - 2) x^{i-j} y^{j-3} \right) c_{ij}^{(2)} \quad (16)$$

Taking $n = \frac{m(m+1)}{2}$ and using a transformation of matrix writing into vector writing we set $\forall k \in \{1, \ldots, n\}$:

$$a_k = x^{i-j} y^{j-1} \quad (17)$$

$$e_k = (i - j) x^{i-j-1} y^{j-1} \cos\theta + (j - 1) x^{i-j} y^{j-2} \sin\theta \quad (18)$$

$$d_k = (i - j)(i - j - 1) x^{i-j-2} y^{j-1} + (j - 1)(j - 2) x^{i-j} y^{j-3} \quad (19)$$

The choice of n_1 points on the boundary $\partial\Omega$ and n_2 collocation points inside the domain allows us to reformulate the problem in matrix form:

$$\mathbf{A_2 c^{(2)}} = \mathbf{b_2} \quad (20)$$

with $\mathbf{b_2} \in \mathbb{R}^{n_c}$, $n_c = 2n_1 + n_2$ and $\mathbf{A_2} \in \mathbb{R}^{n_c \times n}$ given by

$$\mathbf{A_2} = \begin{bmatrix} a_1^T \\ \vdots \\ a_{n_1}^T \\ e_1^T \\ \vdots \\ e_{n_1}^T \\ d_1^T \\ \vdots \\ d_{n_2}^T \end{bmatrix} \text{ and } \mathbf{b_2} = \begin{bmatrix} f(\theta_1) \\ \vdots \\ f(\theta_{n_1}) \\ -\frac{1}{\sigma_1} g(\theta_1) \\ \vdots \\ -\frac{1}{\sigma_1} g(\theta_{n_1}) \\ 0 \\ \vdots \\ 0 \end{bmatrix} \quad (21)$$

Once the approximation of $\mathbf{c}^{(2)}$ is obtained, we reconstruct the approximate solution (12) and its derivative (15).

Then we perform a polynomial expansion of u_1:

$$u_1(x, y) = \sum_{i=1}^{m} \sum_{j=1}^{i} c_{ij}^{(1)} x^{i-j} y^{j-1} \tag{22}$$

And we solve (9) in the same way as (8) by calculating $\mathbf{A_1}$ in the same way as $\mathbf{A_2}$ and taking:

$$\mathbf{b_1} = \begin{bmatrix} u_2(r_1, \theta_1) \\ \vdots \\ u_2(r_1, \theta_{n_1}) \\ -\frac{\sigma_2}{\sigma_1} \partial_\nu u_2(r_1, \theta_1) \\ \vdots \\ -\frac{\sigma_2}{\sigma_1} \partial_\nu u_2(r_1, \theta_{n_1}) \\ 0 \\ \vdots \\ 0 \end{bmatrix} \tag{23}$$

Once the coefficients $c_{ij}^{(1)}$ are obtained, the potential function $u_1(x, y)$ and its normal derivative $\partial_\nu u_1(x, y)$ can be reconstructed using Eqs. (22) and (15), respectively. Finally, the neurocortical potentials can be calculated from the reconstructed potential function using an appropriate forward model.

In summary, the polynomial expansion method provides a semi-analytical approach to solving the Cauchy problem associated with the noninvasive calculation of neurocortical potentials from scalp potentials. By representing the potential function as a polynomial, the method reduces the problem to a set of linear equations, which can be efficiently solved to obtain the coefficients of the polynomial. While the method has its limitations, such as its reliance on simplified head models and assumptions of linearity, it can serve as a valuable tool for investigating the underlying mechanisms of brain activity and informing clinical diagnoses and treatments.

4.3 Solving the Resulting System of Linear Equations for the Expansion Coefficients

After obtaining the system of linear equations for the expansion coefficients, the next step is to solve the system to obtain the coefficients $c_{ij}^{(1)}$ and $c_{ij}^{(2)}$. This can be done by expressing the system in matrix form as:

$$\mathbf{Ac} = \mathbf{b} \tag{24}$$

with $\mathbf{b} \in \mathbb{R}^{n_c}$, $n_c = 2n_1 + n_2$ and $\mathbf{A} \in \mathbb{R}^{n_c \times n}$.

Once the coefficients $c_{ij}^{(1)}$ and $c_{ij}^{(2)}$ have been determined, the potential $u_1(x, y)$ and its normal derivative $\partial_\nu u_1(x, y)$ on Γ_1 and the potential $u_2(x, y)$ and its normal derivative $\partial_\nu u_2(x, y)$ on Γ_2 can be computed using Eqs. (22)–(15).

It should be noted that in order to uniquely determine the solution of the system of linear algebraic equation (24), the number n_c of collocation points and the number n must satisfy the inequality $n_c \geq n$.

Note also that, in general system (24) can not be solved using standard methods of linear algebra, such as Gaussian elimination or LU decomposition. since it is a rectangular system, that is why a least squares method must be used.

Tikhonov's Regularization In the previous section, we derived a system of linear equations for the coefficients c_{ij} that we can use to solve for the potential and its normal derivative in the annular model. However, the system is often ill-conditioned, which means that small errors in the measured scalp potentials can lead to large errors in the estimated coefficients and hence in the reconstructed potential. To overcome this problem, we can add a regularization term to the system of linear equations.

The most commonly used regularization method is Tikhonov regularization, which adds a penalty term to the least-squares problem (the reader may consult [70] for other regularization procedures). The regularized problem can be written as follows:

$$\left[A^T A + \alpha \mathbf{I}\right] C_\alpha = A^T b \tag{25}$$

where α is a positive scalar that controls the amount of regularization, $\mathbf{I}$ is the identity matrix, and A is the same as in the previous section. The penalty term $\alpha\mathbf{I}$ ensures that the solution is smooth and reduces the effects of noise in the data.

The regularized coefficients $\mathbf{c}^{(1)}$ and $\mathbf{c}^{(2)}$ can be found by solving the regularized least-squares problem using a matrix factorization method such as QR decomposition or singular value decomposition (SVD).

The choice of the regularization parameter α is critical, as it determines the trade-off between fitting the data and smoothing the solution. A common approach is to use cross-validation to choose an optimal value of α that minimizes a measure of prediction error on a validation set.

Note that if $\alpha = 0$ this system is equivalent to solving the following normal equation:

$$Dc = b_1, \tag{26}$$

where $b_1 = A^T b$ and $D = A^T A$. D is a symmetric positive definite matrix, thus the conjugate gradient (CG) method can be used to solve this last linear system.

Overall, regularization is a powerful tool for improving the stability and accuracy of the polynomial expansion method for solving the Cauchy problem associated with noninvasive calculation of neurocortical potentials from scalp potentials in the annular model.

Yet, there are times when the contributions of data mistakes and rounding errors will fully outweigh this least squares solution, as we will see in the numerical results. It is feasible to reduce their contributions by including a regularization.

However, a preconditioning is required because the effect of the poor conditioning of these matrices cannot be avoided even by regularization.

Preconditioning Solving the regularized system of linear equations for the coefficients $\mathbf{c}^{(1)}$ and $\mathbf{c}^{(2)}$ using iterative methods such as the conjugate gradient method can be computationally expensive and time-consuming, particularly for large values of m. However, one can use a preconditioning technique to accelerate the convergence of the iterative solver.

The term *preconditioner* refers to a matrix with a suitable advantage that may be used to create a new system that has the same solution as the original system but is better conditioned, or whose number of consequent conditions is reduced. Right, left, and two-sided preconditioning are the types of preconditioning we apply (as defined in [5]).

The left and right preconditioning matrices P and Q are diagonal defined by: $P = diag(p_1, p_2 \cdots p_n)$ and $Q = diag(q_1, q_2 \cdots q_n)$ where p_k and q_k are given by:

$$p_k = \gamma \left(\frac{\sum_{i=1}^{n} A_{i1}^2}{\sum_{i=1}^{n} A_{ik}^2} \right)^{\frac{1}{2}}, \qquad q_k = \delta \left(\frac{\sum_{i=1}^{n} A_{1i}^2}{\sum_{i=1}^{n} A_{ki}^2} \right)^{\frac{1}{2}} \tag{27}$$

for $k = 1, \cdots, n$. The parameters γ and δ are amplifying factors introduced to more reduce the condition number. If $\gamma = \delta = 1$, then the norm of each row and column of the obtained matrices QA and AP are the same respectively.

Overall, the multi-scale preconditioning technique can significantly reduce the computational cost of solving the regularized system of linear equations for the coefficients $c_{ij}^{(2)}$, making it a useful tool for noninvasive calculation of neurocortical potentials from scalp potentials.

Note that, Once the expansion coefficients $\mathbf{c}^{(1)}$ and $\mathbf{c}^{(2)}$ are obtained, we can compute the potential V and the neurocortical potentials ϕ at any point in the brain volume using the lead field matrix and appropriate inverse techniques.

5 Numerical Results

To evaluate the effectiveness of the proposed method, we conducted numerical experiments. In our simulations, we assumed a head model consisting of two concentric annuli representing the brain, skull, and scalp with radii $r_0 = 0.087$m, $r_1 = 0.92$, and $r_2 = 1$m, respectively. We used 100 electrodes placed uniformly on the scalp surface. For the conductivity parameters we set $\sigma_1 = 1$ and $\sigma_2 = \frac{1}{80}$.

In order to compare the approximate solutions to the exact solutions, we solve a more general problem than the Cauchy problem (7), we consider that the jumps defined in (6) are not necessarily null. We therefore provide harmonic exact solutions on Ω_2 and on Ω_1 with a jump on Γ_1:

$$[u_{ex}]_{\Gamma_1} = \Psi(r_1, \theta) \tag{28}$$

$$[\sigma \frac{\partial u_{ex}}{\partial \nu}]_{\Gamma_1} = \Phi(r_1, \theta) \tag{29}$$

Then on Γ_1 we set:

$$u_1 = u_2 + \Psi \tag{30}$$

$$\frac{\partial u_1}{\partial r_1} = -\frac{\sigma_2}{\sigma_1}\frac{\partial u_2}{\partial r_1} + \Phi \tag{31}$$

In order to compare the following approximations to the exact solutions, the RMSE error is calculated by the following formula:

$$Err = \sqrt{\frac{\sum_{i=1}^{N}(u_i - (u_{ex})_i)^2}{N}} \tag{32}$$

where u_i and $u_{ex})_i$ are the values of $u_i^{(1)}$ and of the exact solution calculated respectively at the point (r_0, θ_i) of the border Γ_0.

Example 1 First, we computed data in (7) from the exact solution $u_{\text{ex}}^{(1)} = x^2 - y^2$ using (30)–(31). The number of boundary collocation points used for discretizing the boundary is taken to be $\text{n}_1 = 100$, and for the number of internal collocation points is $n_2 = 1000$ giving $n_c = 1200$.

The best result is obtained for $m = 3$. This is normal, since we approximate a polynomial solution of degree 2 by a polynomial whose degree is $m - 1$. The best polynomial approximating a polynomial of degree q is a polynomial of the same degree. We represent the solutions for $m = 3$ in Fig. 2.

Here, the RMSE error is 3.66×10^{-10}, so it is of the order of machine zero, which is consistent because the exact solution is a polynomial of degree 2.

Example 2 Next, we applied the proposed polynomial expansion method to solve the Cauchy problem and obtain the potential distribution on the surface of the brain region. The given data in (7) are computed $u_{\text{ex}}^{(1)} = (\alpha x + \beta)(\gamma y + \delta)$, With $\alpha = 1$, $\beta = 2$, $\delta = 5$ and $\gamma = 1$ using (30)–(31). The number of internal and boundary collocation points remains the same as in example 1.

We observe from Fig. 3 that the error is very good in the neighborhood of the scalp of the order 10^{-6}, it deteriorates up to 10^{-3} on the skull-scalp interface then

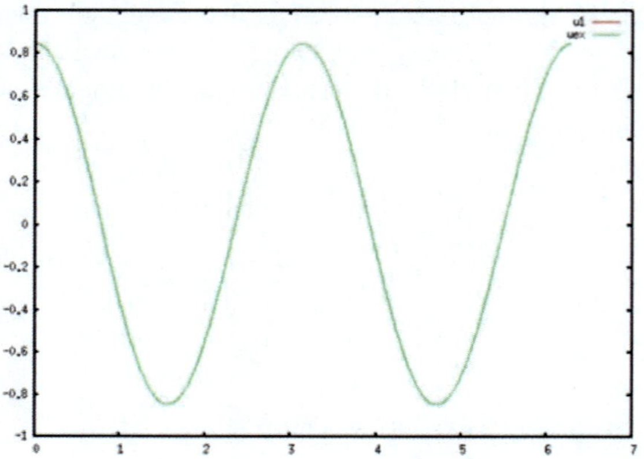

Fig. 2 Exact and computed solution for Example 1

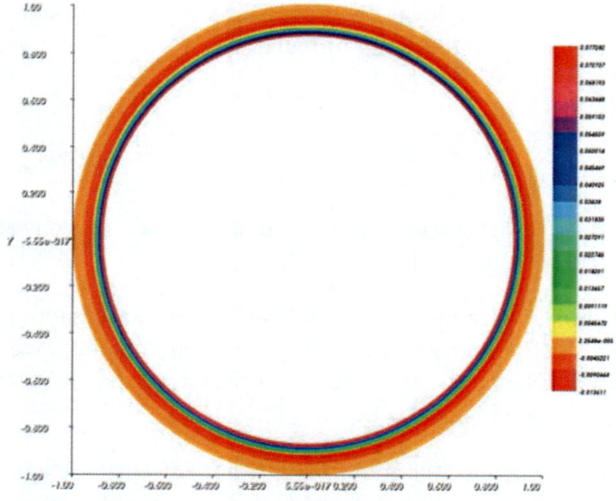

Fig. 3 Difference between approximate and exact solution for Example 2

it returns of the order of 7×10^{-2} on the surface of the brain. This is not due to the polynomial expansion but to the method of successive resolution used. Indeed the Cauchy data for the problem on the skull are not exact as for the first problem on the scalp. The decrease in accuracy is expected in the skull, since, since the potential on the surface of the brain is obtained from the Cauchy conditions on the skull-scalp interface which present an error of order 10^{-3}. Considering that we are solving a noise-sensitive Cauchy problem, we therefore understand that the solution on the second annulus is less suitable than on the first one (Fig. 4).

Example 3 Next, we applied the proposed polynomial expansion method to solve the Cauchy problem and obtain the potential distribution on the surface of the brain

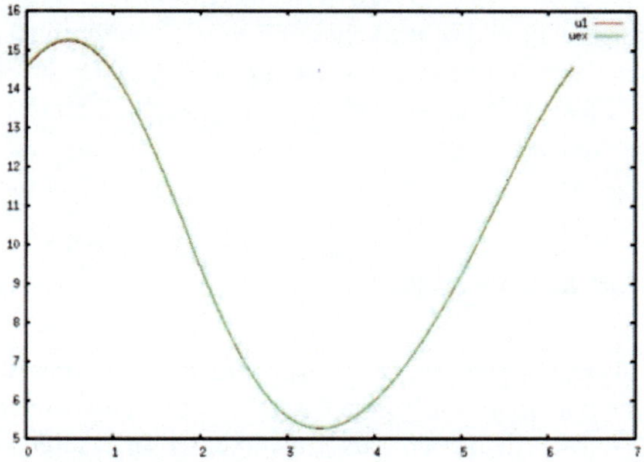

Fig. 4 Exact and computed solution for Example 2

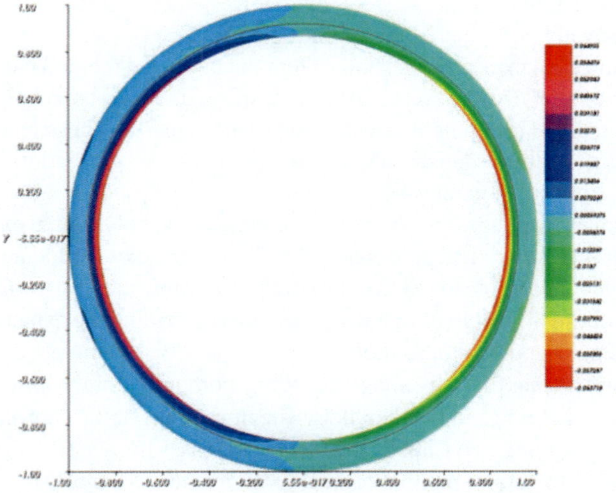

Fig. 5 Difference between approximate and exact solution for Example 3

region. The given data in (7) are computed from the following non-polynomial exact solution $u_{\text{ex}}^{(1)} = u(x, y) = \frac{x}{r^3} = \frac{x}{(\sqrt{x^2+y^2})^3}$. The values of the other given data are kept as in the above examples.

The best approximation with the polynomial expansion is obtained with $m = 5$. Note that here we have no indication to determine m, so we had to do several tests to find the right m.

This time, the error is 6.4×10^{-2} (with preconditioning). The approximation is suitable. We observe from the Fig. 5 that the approximation is better in the annulus representing the scalp and that it deteriorates more and more when approaching the surface of the brain Γ_0. For example the error is of the order of 7×10^{-3} on the skull-scalp interface Γ_1 and of the order of 7×10^{-2} on the surface of the brain Γ_0. This confirms the observations made in example 2. These results and others obtained by varying the different parameters show that the method produces stable solutions with suitable accuracies.

6 Discussion and Conclusion

The objective of this paper is to develop an analytical method based on the polynomial expansion of the potential to solve the Cauchy problem associated with noninvasive calculation of neurocortical potentials from potentials measured at the scalp surface. Specifically, we consider the annular head model and the Laplace equation governing the electroencephalography (EEG) forward problem. We aim to compute the neurocortical potentials from the Cauchy data corresponding to the scalp potentials and normal derivative of the potential at the scalp surface.

Our numerical experiments showed that the proposed method can accurately estimate the potential distribution on the brain surface The average relative error between the estimated and true potential values was less than 7%. These results demonstrate the potential of the proposed method for noninvasive calculation of neurocortical potentials from scalp potentials.

In summary, the sucessive approach combined with polynomial expansion method provides a semi-analytical solution to the Cauchy problem associated with the noninvasive calculation of neurocortical potentials from scalp potentials. The method involves solving a system of linear equations to determine the expansion coefficients of the potential in the annular model.

The method has the advantage of being computationally efficient and can be applied to a wide range of head models. The limitations of the approach are that: first, successively solving two Cauchy problems decreases the accuracy of the solution on the surface of the brain, but this can be circumvented by simultaneously solving the two problems. Then the appropriate degree in the polynomial expansion not being determined for any Cauchy data, it is necessary to take fairly high degrees in order to hope for the production of a good approximation. But this makes the linear systems to be solved very ill-conditioned. We can still obtain correct solutions using the proposed preconditioning procedure.

References

1. C.C. Wood, Human brain mapping in bothe time and space, Haman Brain Map. 3–5 (1994)
2. M. Clerc, J. Kybic, Cortical mapping by laplace-cauchy transmission using a boundary element method. Inverse Prob. **23**(6), 2589 (2007)
3. N. Koshev, N. Yavich, M. Malovichko, E. Skidchenko, M. Fedorov, Fem-based scalp-to-cortex eeg data mapping via the solution of the cauchy problem. J. Inverse Ill-Posed Prob. **28**(4), 517–532 (2020)

4. F. Aboud, A. Nachaoui, M. Nachaoui, On the approximation of a cauchy problem in a non-homogeneous medium. J. Phys.: Conf. Ser. **1743**(1), 012003 (2021)
5. S.M. Rasheed, A. Nachaoui, M.F. Hama, A.K. Jabbar, Regularized and preconditioned conjugate gradient like-methods methods for polynomial approximation of an inverse cauchy problem. Adv. Math. Models Appl. **6**(2), 89–105 (2021)
6. A. Nachaoui, F. Aboud, Solving geometric inverse problems with a polynomial based meshless method, in *New Trends of Mathematical Inverse Problems and Applications* (Springer Proceedings in Mathematics & Statistics, Springer Cham, 2023)
7. S. Rush, D.A. Driscoll, Current distribution in the brain from surface electrodes. Anesth. Analg. 717–723 (1968)
8. J.C. Mosher, Forward solutions for inverse methods. IEEE Trans. Biomed. Eng. 245–259 (1999)
9. R.D. Pascual-Marqui, Review of methods for solving the eeg inverse problem **1**, 75–86 (1999)
10. C.H. Huang, W.C. Chen, A three-dimensional inverse forced convection problem in estimating surface heat flux by conjugate gradient method. Int. J. Heat Mass Transf. **43**(17), 3171–3181 (2000)
11. A. Nachaoui, An improved implementation of an iterative method in boundary identification problems. Numer. Algorithms **33**(1–4), 381–398 (2003)
12. W. Fang, M. Lu, A fast collocation method for an inverse boundary value problem. Int. J. Numer. Methods Eng. **21**, 1563–1585 (2004)
13. A. Nachaoui, Numerical linear algebra for reconstruction inverse problems. J. Comput. Appl. Math. **162**, 147–164 (2004)
14. C.L. Fu, X.L. Feng, Z. Qian, The fourier regularization for solving the cauchy problem for the helmholtz equation. Appl. Numer. Math. **59**(10), 2625–2640 (2009)
15. A. Chakib, A. Nachaoui, A. Zeghal, A shape optimization approach for an inverse heat source problem. Int. J. Nonlinear Sci. **12**(1), 78–84 (2012)
16. A. Chakib, A. Ellabib, A. Nachaoui, M. Nachaoui, A shape optimization formulation of weld pool determination. Appl. Math. Lett. **25**(3), 374–379 (2012)
17. A. Boulkhemair, A. Nachaoui, A. Chakib, A shape optimization approach for a class of free boundary problems of bernoulli type. Appl. Math. **58**(2), 205–221 (2013)
18. M.M. Lavrentiev, *Some Improperly Posed Problems of Mathematical Physics*. (Springer Science & Business Media, 2013)
19. H.F. Guliyev, Y.S. Gasimov, S.M. Zeynalli, Optimal control method for solving the cauchy-neumann problem for the poisson equation. Zh. Mat. Fiz. Anal. Geom. **12**(4), 412–421 (2014)
20. A. Chakib, M. Johri, A. Nachaoui, M. Nachaoui, On a numerical approximation of a highly nonlinear parabolic inverse problem in hydrology. An. Univ. Craiova Ser. Mat. Inform. **42**, 192–201 (2015)
21. L. Rincon, S. Shimoda, The inverse problem in electroencephalography using the bidomain model of electrical activity. J. Neurosci. Methods **274**, 94–105 (2016)
22. V. Isakov, *Inverse Problems for Partial Differential Equations*, vol. 127 of Applied Mathematical Sciences (Springer Cham, 2017)
23. C.S. Liu, F. Wang, A meshless method for solving the nonlinear inverse cauchy problem of elliptic type equation in a doubly-connected domain. Comput. Math. Appl. **76**, 1831–1852 (2018)
24. A. Bergam, A. Chakib, A. Nachaoui, M. Nachaoui, Adaptive mesh techniques based on a posteriori error estimates for an inverse cauchy problem. Appl. Math. Comput. **346**, 865–878 (2019)
25. A. Chakib, A. Hadri, A. Laghrib, On a multiscale analysis of an inverse problem of nonlinear transfer law identification in periodic microstructure. An. Univ. Craiova Ser. Mat. Inform. **51**, 102985 (2020)
26. F. Wang, Y. Gu, W. Qu, C. Zhang, Localized boundary knot method and its application to large-scale acoustic problems. Comput. Methods Appl. Mech. Eng. **361**, 112729 (2020)
27. A. Nachaoui, M. Nachaoui, A. Chakib, M. Hilal, Some novel numerical techniques for an inverse cauchy problem, J. Comput. Appl. Math. **381**(113030)

28. A. Nachaoui, A. Laghrib, M. Hakim, *Mathematical Control and Numerical Applications*, vol. 372 of Springer Proc. Math. Stat (Springer, Cham, 2021)
29. M. Nachaoui, A. Nachaoui, T. Tadumadze, On the numerical approximation of some inverse problems governed by nonlinear delay differential equation. RAIRO Oper. Res. **56**, 1553–1569 (2022)
30. L. Afraites, C. Masnaoui, M. Nachaoui, Shape optimization method for an inverse geometric source problem and stability at critical shape. Discrete Contin. Dyn. Syst. Ser. S **15**(1), 1–21 (2022)
31. H. Ouaissa, A. Chakib, A. Nachaoui, M. Nachaoui, On numerical approaches for solving an inverse cauchy stokes problem. Appl. Math. Optim. **85**(1). https://doi.org/10.1007/s00245-022-09833-8
32. A. Ellabib, A. Nachaoui, A. Ousaadane, Convergence study and regularizing property of a modified robin-robin method for the cauchy problem in linear elasticity. Inverse Prob. **38**, 075007 (2022)
33. K. Berdawood, A. Nachaoui, M. Nachaoui, An accelerated alternating iterative algorithm for data completion problems connected with helmholtz equationn. Stat. Optim. Inf. Comput. **11**(1), 2–21 (2023). https://doi.org/10.1002/num.22793
34. M. Clerc, J. Kybic, Basic mathematical and electromagnetic concepts of the biomagnetic inverse problem. Phys. Med. Biol. **32**(1), 11–22 (1987)
35. A.R. Johnson, R.S. MacLeod, Adaptive local regularization methods for the inverse ecg problem. Prog. Biophys. Mol. Biol. **69**(2–3), 405–423 (1998)
36. R. Grech, T. Cassar, J. Muscat, K.P. Camilleri, S.G. Fabri, M. Zervakis, P. Xanthopoulos, V. Sakkalis, B. Vanrumste, Review on solving the inverse problem in eeg source analysis. J. NeuroEng. Rehabil. **5**(1), 25 (2008)
37. T.V. Zakharova, P.I. Karpov, V.M. Bugaevskii, Localization of the activity source in the inverse problem of magnetoencefalography. Comput. Math. Model. **28**(2), 148–157 (2017)
38. E. Hernandez-Montero, A. Fraguela-Collar, J. Henry, An optimal quasi solution for the cauchy problem for laplace equation in the framework of inverse ecg. Math. Model. Nat. Phenom. **14**
39. M. Hämäläinen, R. Hari, R.J. Ilmoniemi, J. Knuutila, O.V. Lounasmaa, Magnetoencephalography: theory, instrumentation, and applications to noninvasive studies of the working human brain. Rev. Modern Phys
40. F. Aboud, I.T. Jameel, A.F. Hasan, B.K. Mostafa, A. Nachaoui, Polynomial approximation of an inverse cauchy problem for helmholtz type equations. Adv. Math. Models Appl. **7**(3), 306–322 (2022)
41. J. Blum, *Numerical Simulation and Optimal Control in Plasma Physics with Applications to Tokamaks* (Wiley, Chichester, 1989)
42. J.C. Liu, T. Wei, A quasi-reversibility regularization method for an inverse heat conduction problem without initial data. Appl. Math. Comput. **219**(23), 10866–10881 (2013)
43. M. Malovichko, N. Koshev, N. Yavich, A. Razorenova, M. Fedorov, Electroencephalographic source reconstruction by the finite-element approximation of the elliptic Cauchy problem. EEE Trans. Biomed. Eng. **68**(6), 1811–1019 (2020)
44. D. Maxwell, M. Truffer, S. Avdonin, M. Stueferv, An iterative scheme for determining glacier velocities and stresses. J. Glaciol. **54**(188), 888–898 (2008)
45. A. Nachaoui, S.M. Rashid, A mesh free wavelet method to solve the cauchy problem for the helmholtz equation, in *New Trends of Mathematical Inverse Problems and Applications* (Springer Proceedings in Mathematics & Statistics, Springer Cham, 2023)
46. P. Dvalishvili, A. Nachaoui, M. Nachaoui, T. Tadumadze, On the well-posedness of the cauchy problem for a class of differential equations with distributed delay and the continuous initial condition. Proc. Inst. Math. Mech. Natl. Acad. Sci. Azerb. **43**, 146–160 (2017)
47. J. Hadamard, *Lectures on Cauchy's Problem in Linear Partial Differential Equations* (Yale University Press, New Haven, 1923)
48. M.M. Lavrentiev, *Some Improperly Posed Problems of Mathematical Physics* (SIAM, PhiladelphiaSpringer Science and Business Media, 2013)

49. M. Jourhmane, A. Nachaoui, A relaxation algorithm for solving a cauchy problem, in *Proceedings of the Second International Conferences on Inverse Problems in Engineering* (Engineering Foundation, 1996), pp. 151–158
50. M. Jourhmane, A. Nachaoui, Convergence of an alternating method to solve the cauchy problem for poisson's equation. Appl. Anal. **81**(5), 1065–1083 (2002)
51. A. Nachaoui, F. Aboud, M. Nachaoui, Acceleration of the kmf algorithm convergence to solve the cauchy problem for poisson's equation, in *Mathematical Control and Numerical Applications*, vol. 372 of Springer Proceedings in Mathematics & Statistics, eds. by A. Nachaoui, A. Hakim, A. Laghrib (Springer, Cham, 2021), pp. 43–57
52. M. Essaouini, A. Nachaoui, S. El Hajji, Numerical method for solving a class of nonlinear elliptic inverse problems. J. Comput. Appl. Math. **162**(1), 165–181 (2004)
53. M. Essaouini, A. Nachaoui, S. El Hajji, Reconstruction of boundary data for a class of nonlinear inverse problems. J. Inverse Ill-Posed Prob. **12**(4), 369–385 (2004)
54. A. Ellabib, A. Nachaoui, An iterative approach to the solution of an inverse problem in linear elasticity. Math. Comput. Simul. **77**, 189–201 (2008)
55. L. Marin, B.T. Johansson, A relaxation method of an alternating iterative algorithm for the cauchy problem in linear isotropic elasticity. Comput. Methods Appl. Mech. Eng. **199**(49–52), 3179–3196 (2010)
56. A. Ellabib, A. Nachaoui, A. Ousaadane, Mathematical analysis and simulation of fixed point formulation of cauchy problem in linear elasticity. Math. Comput. Simul. **187**, 231–247 (2021)
57. K.A. Berdawood, A. Nachaoui, M. Nachaoui, F. Aboud, An effective relaxed alternating procedure for cauchy problem connected with helmholtz equation. Numer. Methods Partial Differ. Equ. 1–27 (2021)
58. K. Berdawood, A. Nachaoui, R. Saeed, M. Nachaoui, F. Aboud, An efficient dn alternating algorithm for solving an inverse problem for helmholtz equation. Discrete Contin. Dyn. Syst. Ser. S
59. A. Chakib, A. Nachaoui, M. Nachaoui, H. Ouaissa, On a fixed point study of an inverse problem governed by stokes equation. Inverse Prob. **35**, 015008 (2019)
60. A. Nachaoui, Iterative methods for inverse problems subject to the convection-diffusion equation, in *New Trends of Mathematical Inverse Problems and Applications* (Springer Proceedings in Mathematics & Statistics, Springer, Cham, 2023)
61. S. Yarmukhamedov, I. Yarmukhamedov, Cauchy problem for the helmholtz equation, in *Ill-Posed and Non-Classical Problems of Mathematical Physics and Analysis*, Inverse Ill-posed Probl. Ser. (VSP, Utrecht, 2003), pp. 143–172
62. A. Arsenashvili, A. Nachaoui, T. Tadumadze, On approximate solution of an inverse problem for linear delay differential equations. Bull. Georgian Natl. Acad. Sci. (N.S.) **2**(2), 24–28 (2008)
63. R. Shi, T. Wei, H.H. Qin, A fourth-order modified method for the Cauchy problem of the modified Helmholtz equation. Numer. Math. Theory Methods Appl. **2**(3), 326–340 (2009)
64. A.L. Qian, X.T. Xiong, Y.-J. Wu, On a quasi-reversibility regularization method for a cauchy problem of the helmholtz equation. J. Comput. Appl. Math. **233**(8), 1969–1979 (2010)
65. S.I. Kabanikhin, *Inverse and ill-posed Problems* (Walter de Gruyter GmbH & Co. KG, Berlin, 2012)
66. Q. Hua, Y. Gu, W. Qu, W. Chen, C. Zhang, A meshless generalized finite difference method for inverse cauchy problems associated with three-dimensional inhomogeneous helmholtz-type equations. Eng. Anal. Bound. Elem. **82**, 162–171 (2017)
67. Z. Qian, X. Feng, A fractional tikhonov method for solving a cauchy problem of helmholtz equation. Appl. Anal. **96**, 1656–1668 (2017)
68. K.A. Berdawood, A. Nachaoui, R. Saeed, M. Nachaoui, F. Aboud, An alternating procedure with dynamic relaxation for cauchy problems governed by the modified helmholtz equation. Adv. Math. Models Appl. **5**(1), 131–139 (2020)
69. A. Nachaoui, H.W. Salih, An analytical solution for the nonlinear inverse cauchy problem. Adv. Math. Models Appl. **6**(3), 191–206 (2021)
70. P.C. Hansen, *Rank-Deficient and Discrete ill-posed Problems: Numerical Aspects of Linear Inversion* (SIAM, Philadelphia, 1998)

Solving Geometric Inverse Problems with a Polynomial Based Meshless Method

Abdeljalil Nachaoui and Fatima Aboud

Abstract In this paper, we present a novel method for solving an inverse problem that involves determining an unknown defect D compactly contained in a simply-connected bounded domain Ω, given the Dirichlet temperature data u, the Neumann heat flux data $\partial_\nu u$ on the boundary $\partial\Omega$, and a Dirichlet boundary condition on the boundary ∂D. We assume that the temperature u satisfies the modified Helmholtz equation governing the conduction of heat in a fin. The proposed method involves dividing the problem into two subproblems: first, solving a Cauchy problem governed by the modified Helmholtz equation to determine the temperature u, followed by solving a series of nonlinear scalar equations to determine the coordinates of the points defining the boundary ∂D. Our numerical experiments demonstrate the effectiveness and accuracy of the proposed method in solving this challenging inverse problem.

Keywords Modified Helmholtz's equation · Inverse problem · Cauchy problem · Polynomial expansion · Regularisztion

1 Introduction

In many physical applications, it is important to determine the properties of an object or material based on measurements of temperature and heat flux. One common problem is the inverse problem of determining the location and shape of an unknown defect in a bounded domain from these measurements.

In this paper, we present a method for solving this inverse problem assuming that the temperature satisfies the modified Helmholtz equation governing the conduction

---

A. Nachaoui (✉)
Laboratoire de Mathématiques Jean Leray, Nantes Université, Nantes, France
e-mail: Abdeljalil.Nachaoui@univ-nantes.fr

F. Aboud
Department of Mathematics, College of Science, University of Diyala, Baqubah, Iraq
e-mail: Fatima.Aboud@uodiyala.edu.iq

A. Laghrib et al. (eds.), *New Trends of Mathematical Inverse Problems and Applications*, Springer Proceedings in Mathematics & Statistics 428,
https://doi.org/10.1007/978-3-031-33069-8_8

of heat in a fin. The proposed method involves dividing the problem into two subproblems. In the first subproblem, we solve a Cauchy problem governed by the modified Helmholtz equation to determine the temperature u. In the second subproblem, we solve a series of nonlinear scalar equations to determine the coordinates of the points defining the boundary ∂D.

The Cauchy problem, is one of the examples of inverse problems [1–21].

For the Cauchy problem governed by the Poisson equation the authors [22–24] have proposed the relaxation of the Dirichlet data given in the case of the alternating iterative algorithm proposed in [25]. This procedure considerably reduced the number of iterations needed to achieve convergence of the considered inverse problems. Recently, this acceleration result was proved in [26] for poisson equation and in [27] for the Helmholtz equation. This technique was also used for a nonlinear elliptical problem [28] and [29], then in structural mechanics [30–32], in electromagnetism [33] and in fluid mechanics where the method has been applied to the Cauchy problem governed by Stocks equation [34].

Other methods have been developed to solve this kind of problem, the reader can consult for example [8, 10, 33, 35–43] and the references therein.

With regard to solving the Cauchy problem for the Helmholtz equation, several methods have been proposed over the past two decades [3, 5, 15, 36, 40, 44–53]. Here, by solving the Cauchy problem, we will follow the approach used in [53] based on a polynomial expansion approximated by collocation.

The rest of the paper is organized as follows. In Sect. 2, we provide a mathematical formulation of the inverse problem and describe the assumptions we make. In Sect. 3, we describe our method for solving the inverse problem, including the two subproblems. In Sect. 4, we present numerical experiments to demonstrate the effectiveness and accuracy of the proposed method. The comparison with other results of previous works, in particular [54], shows the simplicity as well as the efficiency of the proposed method. Finally, we conclude the paper in Sect. 5 with a discussion of our results and suggestions for future work.

2 Mathematical Formulation of the Inverse Problem

We consider a bounded domain Ω in $\mathbb{R}^2$ that is simply-connected and has a smooth boundary $\partial\Omega$. Let D be an unknown defect compactly contained in Ω with boundary ∂D, which is assumed to be smooth and piecewise analytic and such that $\Omega_a = \Omega \setminus D$ is connected. We assume that the temperature $u(x, y)$ in Ω_a satisfies the modified Helmholtz equation

$$\Delta u(x, y) - k^2 u(x, y) = F(x, y), \quad (x, y) \in \Omega_a \tag{1}$$

where the wavenumber k is a function of the convective heat transfer coefficient, the thermal conductivity and fin thickness.

We are given the Dirichlet temperature data $u|_{\partial\Omega_a}$, i.e. Dirichlet boundary conditions on the boundary $\partial\Omega$ and ∂D. The steady state temperature u in a fin satisfies the following problem

$$\Delta u(x, y) - k^2 u(x, y) = F(x, y), \quad (x, y) \in \Omega_a \tag{2}$$

$$u(\rho, \theta) = f(\rho, \theta), \quad (\rho, \theta) \in \partial\Omega \tag{3}$$

$$u(\rho, \theta) = u_0(\rho, \theta), \quad (\rho, \theta) \in \partial D \tag{4}$$

where F, f and u_0 are given functions.

It is well-known that the direct Dirichlet problem given by Eqs. (2)–(4) has a unique solution $u \in H^1(\Omega \setminus D)$, when D is known, $F \in L^2(\Omega)$, $f \in H^{1/2}(\partial\Omega)$ and $u_0 \in H^{1/2}(\partial D)$. We can then define a nonlinear operator $G(\partial D)$, which maps from the set of admissible Lipschitz boundaries ∂D to the data space of Neumann heat flux data in $H^{-1/2}(\partial\Omega)$, as follows:

$$G(\partial D) = \partial_\nu u(\rho, \theta) = g(\rho, \theta), \quad \partial\Omega, \tag{5}$$

where $\partial\Omega$ is the accessible part of the boundary and $\partial_\nu u|_{\partial\Omega}$ is the Neumann heat flux data on the boundary $\partial\Omega$.

Then the inverse problem under consideration consists of extracting some information about the boundary ∂D from the data $g = G(\partial D)$. The data (5) may also be only partial, i.e. the flux being measured on a non-zero measure portion $\Gamma \subset \partial\Omega$, rather than on the entire boundary $\partial\Omega$. It is well-known that the inverse problem is nonlinear and ill-posed as opposed to the direct problem which is linear and well-posed. Although one can show, see [10] , that the solution of the inverse problem given by Eqs. (2)–(4) and (5) is unique, this solution does not depend continuously upon errors in the input Cauchy data (3) and (5).

In [54], the author has proposed the meshless method of fundamental solutions (MFS) to approximate u where the coefficients and the radii giving ∂D is determined by imposing the boundary conditions (3)–(4) and (5) in a least-squares sense which recasts into a nonlinear unconstrained minimisation of the following regularised objective function.

$$T(u, r) := \|u - f\|^2_{L^2(\partial\Omega)} + \lambda_1 \|\frac{\partial u}{\partial n} - g\|^2_{L^2(\partial\Omega)} + \lambda_2 \|u - u_0\|^2_{L^2(\partial D)} \tag{6}$$

The method depends on two regularisation parameters, but as cited [54] the choice of one single parameter is already highly nontrivial, and the choice of two or three parameters can be very expensive and difficult to justify.

In the following we present our method.

3 Decoupled Cauchy-Newton Algorithm

Consider the problem where one has to find D from solving (2–5). To solve this inverse problem, we first solve a Cauchy problem governed by the modified Helmholtz equation to determine the temperature u in Ω. We use a polynomial expansion approach, which is well-suited for problems with smooth boundaries. This approach allows us to obtain a solution for u that is given by a polynomial expansion. Once we have obtained a solution for u, we proceed to the second subproblem, which involves determining the coordinates of the points defining the boundary ∂D. We use a series of nonlinear scalar equations to achieve this. We first transform the problem into a parametric form, and then use the expression for u to obtain the non linear equations. We then solve these equations numerically using an iterative method. Thus the proposed method can be summarized in the following algorithm:

Algorithm 1: Decoupled Cauchy-Newton algorithm

1: Solve the Cauchy problem obtained by grouping Eqs. (2), (3) and (5).
2: Integrate the solution of this Cauchy problem into Eq. (4)
3: For $\theta \in [0, 2\pi]$, numerically solve the resulting nonlinear scalar equation using an iterative method such as Newton's.

That is to say that our problem of reconstruction of the domain D is reduced to the resolution of a Cauchy problem governed by the modified Helmholtz equation allowing to determine the temperature u followed by a series of nonlinear scalar equations to determine the coordinates of the points defining the boundary ∂D.

3.1 Polynomial Expansion Approximation of the Cauchy Problem

To approximate the solution of the Cauchy problem (2)–(5), we use a polynomial expansion similar to the one described in [53, 55]. Specifically, we seek a polynomial approximation of the form

$$u(x, y) = \sum_{i=1}^{m} \sum_{j=1}^{i} c_{ij} x^{i-j} y^{j-1} \tag{7}$$

where, c_{ij}, $1 \leq i \leq m$ and $1 \leq j \leq i$ are unknown coefficients to be determined, the number of these unknown is $\frac{m(m+1)}{2}$. Substituting this expansion into the Helmholtz equation (2), we get

$$\sum_{i=1}^{m} \sum_{j=1}^{i} \left(\frac{\partial^2}{\partial x^2} + \frac{\partial^2}{\partial y^2} - k^2 \right) c_{ij} x^{i-j} y^{j-1} = F(x, y). \tag{8}$$

Using the fact that

$$\frac{\partial^2}{\partial x^2}x^n = n(n-1)x^{n-2}$$

and

$$\frac{\partial^2}{\partial y^2}y^{m-1} = (m-1)(m-2)y^{m-3},$$

we can simplify the above equation to

$$\sum_{i=1}^{m}\sum_{j=1}^{i}\left(\frac{(i-j)(i-j-1)}{x^2}+\frac{(j-1)(j-2)}{y^2}-k^2\right)c_{ij}x^{i-j}y^{j-1} = F(x,y). \tag{9}$$

Since this equation must hold for all $(x, y) \in \Omega$, we obtain a system of linear equations for the coefficients c_{ij}, at n_{1b} points of Ω gives rise to a linear system

$$\sum_{i=1}^{m}\sum_{j=1}^{i}\left(\frac{(i-j)(i-j-1)}{x_l^2}+\frac{(j-1)(j-2)}{y_l^2}-k^2\right)c_{ij}x_l^{i-j}y_l^{j-1}$$
$$= F(x_l, y_l) \text{ for } l = 1, \ldots n_{1b} \tag{10}$$

To complete this system, we need to specify boundary conditions for u. Since we are given Dirichlet and Neumann boundary data, we can use these to determine the coefficients c_{ij}. Specifically, note that the normal derivative of $u(x, y)$ can be expressed in the following form (see [56]),

$$\partial_n u = \eta(\theta)\left[cos(\theta)-\frac{\rho'}{\rho^2}sin(\theta)\right]\partial_x u + \eta(\theta)\left[sin(\theta)-\frac{\rho'}{\rho^2}cos(\theta)\right]\partial_y u \tag{11}$$

where $\eta(\theta)$ is given by

$$\eta(\theta) = \frac{\rho(\theta)}{\sqrt{\rho^2(\theta)+\left[\rho'(\theta)\right]^2}} \tag{12}$$

Thus, replacing in (3) the solution u by its approximation given in (7) and using the formula (11) for the normal derivative in the boundary condition (5), gives rise to a system of equations whose evaluation at n_{1a} points of $\partial\Omega$ results in a system of $2 * n_{1a}$ linear equations. Combining this system with the n_{1b} equations obtained from (10), forms a system of linear equations of the form:

$$\mathbf{Ac} = \mathbf{b} \tag{13}$$

where the vector **c**, in which the coefficients to be determined c_{ij} have been reordered, is of length $n_2 = m(m+1)/2$. The known data vector **b** has length $n_1 = 2 * n_{1a} + n_{1b}$ and **A**, the rectangular matrix of the system , is of order $n_1 \times n_2$.

Therefore, solving the inverse Cauchy problem (2)–(3) and (5) is reduced to solving the system of linear algebraic equations (13).

It should be noted that in order to uniquely determine the solution of the system of linear algebraic equation (13), the number n_1 of collocation points and the number n_2 must satisfy the inequality $n_1 \geq n_2$.

Once the coefficients c_{ij} are determined, we can use the polynomial expansion (7) to evaluate the solution $u(x, y)$ of the Cauchy problem (2)–(5) at any point in the domain Ω.

Note that the accuracy of the polynomial approximation depends on the number of terms m in the expansion (7). Increasing m improves the accuracy, but also increases the number of coefficients c_{ij} that need to be determined, and hence the computational cost. Increasing m increases also the degrees of the monomials in the evaluation of the coefficients of the matrix. This increases the ill-conditioned character of the system to be solved. A special treatment must then be applied to obtain suitable solutions. The choice of m should therefore be a balance between accuracy and computational efficiency.

3.2 Regularization and Preconditioning

To solve the linear system of equations obtained from the polynomial approximation of the solution of the Cauchy problem, we need to regularize and precondition the system. To regularize the linear subsystem obtained from the approximation of the Cauchy problem by the polynomial expansion, one can use a Tikhonov regularization method. This method adds a regularization term to the equation to stabilize the solution and reduce the effect of noise in the data. The Tikhonov regularization method involves minimizing the following functional:

$$\Phi(\mathbf{c}) = \frac{1}{2}\|\mathbf{A}\mathbf{c} - \mathbf{b}\|^2 + \frac{\lambda}{2}\|\mathbf{c}\|^2, \tag{14}$$

where **A** is the matrix representing the linear system obtained from the polynomial expansion, **b** is the data vector, **c** is the vector of coefficients in the polynomial expansion, and the regularization parameter λ is a small positive constant. The regularization parameter controls the trade-off between the fidelity to the data and the smoothness of the solution. Formally, the Tikhonov regularized solution of problem (13) is obtained as the solution of the regularized system

$$\left(\mathbf{A}^T\mathbf{A} + \lambda\mathbf{I}\right)\mathbf{c} = \mathbf{A}^T\mathbf{b}, \tag{15}$$

where **I** is the identity matrix.

If this is necessary, i.e. in the case where the regularization does not make it possible to obtain a good approximation or the resolution of the regularized system is very slow, we precondition the regularized system to accelerate the convergence of the iterative solver. One can use as preconditioner an incomplete Cholesky factorization (ICF) preconditioner (see [57]) and the references therein. The ICF preconditioner is used to reduce the number of iterations required to solve the linear system by providing a better initial guess for the iterative solver. The ICF preconditioner involves computing an incomplete Cholesky factorization of the matrix $K^T K + \lambda I$. The factorization is incomplete because only a subset of the entries in the Cholesky factor are computed, resulting in a more efficient and approximate factorization.

Here we use a diagonal preconditioner $\mathbf{M}$, where each coefficient is obtained from the coefficients of the matrix $\mathbf{A}$ as in [55]. We opted for this diagonal preconditioner for its simplicity and efficiency demonstrate in [53, 58]

The preconditioned system is given by

$$\mathbf{M}^{-1}\left(\mathbf{A}^T\mathbf{A} + \lambda\mathbf{I}\right)\mathbf{c} = \mathbf{M}^{-1}\mathbf{A}^T\mathbf{b}. \tag{16}$$

Note that if $\alpha = 0$ this system is equivalent to solving the following normal equation:

$$\mathbf{Dc} = \mathbf{b}_1, \tag{17}$$

where $\mathbf{b}_1 = \mathbf{A}^T b$ and $\mathbf{D} = \mathbf{A}^T\mathbf{A}$. System (15) or its preconditioned one (16) can be solved using a Krylov subspace iterative method such as conjugate gradient (CG).

The regularizing and preconditioning steps help to improve the stability and accuracy of the method by reducing the effects of noise in the data and providing a more efficient solver for the linear system.

3.3 Determining the Unknown Boundary with Newton Method

Suppose now that the Cauchy problem has been solved, that is to say that we have obtained the vector c allowing to value $u(x, y)$ at any point $(x, y) \in \Omega$ or in polar coordinates $0 \leq \rho \leq \rho_0(\theta)$ for $0 \leq \theta \leq 2\pi$ and $0 \leq \rho \leq \rho_0(\theta)$, where $\rho_0(\theta)$ is the curve parametrizing the boundary $\partial\Omega$. We can say that for fixed θ, there exists a $\rho(\theta) \in (0, \rho_0(\theta))$ such as $\rho(\theta) \in \partial\Omega$. Thus, to determine the unknown boundary ∂D, we proceed as follows: for fixed $\theta \in [0, 2\pi]$, using the boundary condition (4) on D we can write at point $(\rho(\theta), \theta) \in \partial D$

$$u(\rho, \theta) = u_0(\rho, \theta), \tag{18}$$

where u is the solution of the Cauchy problem defined by (2)–(3) and (5). we obtain

$$Fr_N(\rho(\theta)) = \sum_{i=1}^{m}\sum_{j=1}^{i} c_{ij} \ (\rho(\theta))^{i-1}(cos(\theta))^{i-j}(sin(\theta))^{j-1} - u_0(\rho(\theta), \theta) = 0 \tag{19}$$

This is a a scalar nonlinear equation in $\rho(\theta)$, thus to find the approximate value of $(\rho(\theta), \theta) \in \partial D$ on which the boundary condition (4) is satisfied, one needs to find the root $\rho(\theta)$ of the nonlinear operator Fr_N defined by the nonlinear equation (19).

Since the Fr_N operator is explicitly defined in terms of $\rho(\theta)$, Newton's method can be used to determine the root of Fr_N for each fixed $\theta \in [0, 2\pi]$.

Note that

$$d_\rho Fr_N(\rho(\theta))=\sum_{i=1}^{m}\sum_{j=1}^{i} c_{ij}(i-1)(\rho(\theta))^{i-2}\,(cos(\theta))^{i-j}(sin(\theta))^{j-1} - d_\rho u_0(\rho(\theta, \theta)) \tag{20}$$

where $d_\rho Fr_N$ is the derivative of the function Fr_N as a function of ρ.

We then build a sequence ρ_n, using Newton's algorithm, starting from an initial value $\rho_0 = \rho_0(\theta)$ and updating by

$$\rho_{n+1} = \rho_n - \frac{Fr_N(\rho_n)}{d_\rho Fr_N(\rho_n)}. \tag{21}$$

We can therefore summarize our procedure to solve the inverse geometric problem for the modified Helmholtz equation (2)–(5) which consists of determining an unknown inner boundary of an annular domain from a pair of boundary Cauchy data (boundary temperature and heat flux) in the following algorithm:

Algorithm 2: Discrete decoupled Cauchy-Newton algorithm

1: Construct the matrix A and the second member b of the linear system (13) using the approximation (7).
2: Solve the linear system (13) by using the regularization (15) and possibly the preconditioning (TSkMP) to obtain the coefficients c_{ij}.
3: For $\theta \in [0, 2\pi]$, choose the initial guess ρ_0 to find the root of the non linear equation (19) using Newton's iteration (21)

4 Numerical Results and Discussion

In this section, we discuss the numerical results obtained using the polynomial approximation described in Sect. 3.1 to solve the Cauchy problem for the modified Helmholtz equation in an annular two-dimensional bounded domain. The obtained linear system in (13) is solved using the CGLS methods. To illustrate our proposed method, we compare our results with those found in [54]. So we study some examples.

Example 1 We start with the first example where we consider a simple two-dimensional detection of an unknown circular (rigid inclusion) inner boundary $D = B(0; r_0)$ of radius $r_0 = 0.7$ within the unit circle $\Omega = B(0; 1)$. We take

$$u(r(\theta), \theta) = u_0(\theta) = e^{0.7(cos(\theta)+sin(\theta))} \text{ on } \partial D$$

and

$$u(1, \theta) = f(\theta) = e^{cos(\theta)+sin(\theta)} \text{ on } \partial\Omega,$$

as the Dirichlet data given by Eqs. (3) and (5), respectively. We also take $k = \sqrt{2}$. We take the normal derivative on $\partial\Omega$ given by

$$\frac{\partial u}{\partial n}(1, \theta) = g(\theta) = (cos(\theta) + sin(\theta))e^{cos(\theta)+sin(\theta)}.$$

For this input data, the analytical solution is

$$u(r, \theta) = e^{r(cos(\theta)+sin(\theta))}.$$

The initial guess for the inner boundary is arbitrary and is taken, say to be an ellipse of semi-axes 0.5 and 0.4. For the decoupled Cauchy-Newton algorithm, we take $n_{1a} = 20$, $n_{1b} = 100$ and for stopping criterion in the Newton iteration we take $|\rho^{n+1} - \rho^n| \leq Tol = 10^{-6}$. The results taken from [54] are computed with a number of collocation $M = N = 20$.

In Fig. 1, we present results obtained for Example 1, (Fig. 1a) with decoupled Cauchy-Newton algorithm, (Fig. 1a) results from [54]. From this figure it can be

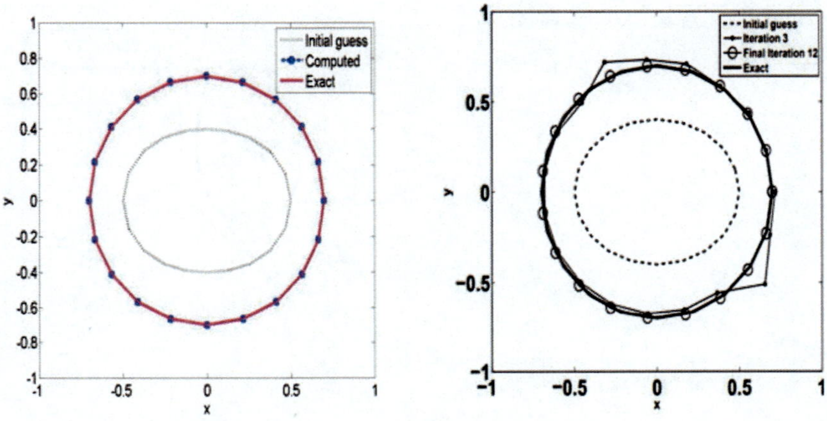

Decoupled Cauchy-Newton algorithm Minimisation of the functional with MFS

Fig. 1 The reconstructed boundary for Example 1 without noise, without regularisation seen that the circular inclusion is accurately identified for both cases. Note that the minimization method requires 12 iterations to accurately identify circular inclusion, i.e. the resolution of several linear systems while our method requires the resolution of a linear system and N nonlinear equation which only require polynomial function evaluations.

5 Noise Effect

In real life problem in which inverse problems are stated, the known data are collected by using measurements for which some error can produced. In order to investigate the stability of the numerical solution, we add noise into the Neumann boundary data g given by expression (5).

The following formula is used for perturbing the exact Cauchy data:

$$g(\theta) = \partial_n u_{ex}(\rho(\theta), \theta) + \sigma * \omega \tag{22}$$

where ω is the random distributed error of Gauss and σ is the standard deviation of measurement errors. σ represent the noise level, with the values $0.001, 0.01, 0.05$ and 0.1.

Example 2 In this example, all the given data in (2)–(3) were taken to be the same as in Example 1, only the boundary condition on the normal derivative (5) is modified as in the (22) equation.

Figure 2 shows the results obtained for Example 2 with a level of noise $\sigma = 1\%$. It can be seen, From Fig. 2b, that the inner boundary solution produced when

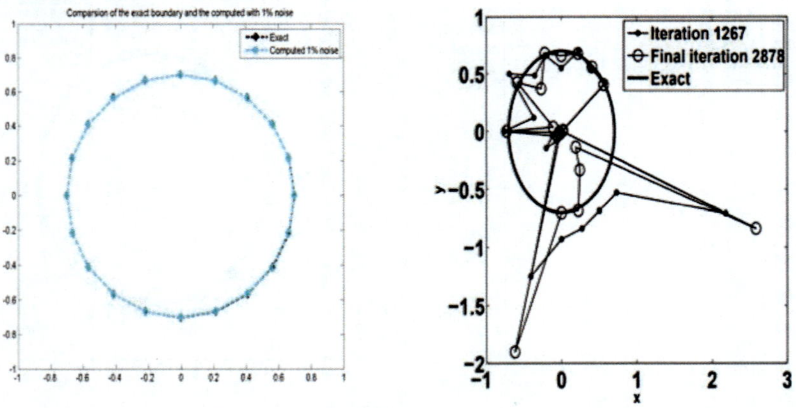

Decoupled Cauchy-Newton algorithm without regularisation

Minimisation of the functional at various iterations without regularisation.

Fig. 2 The reconstructed boundary for Example 2 for a level of noise $\sigma = 1\%$

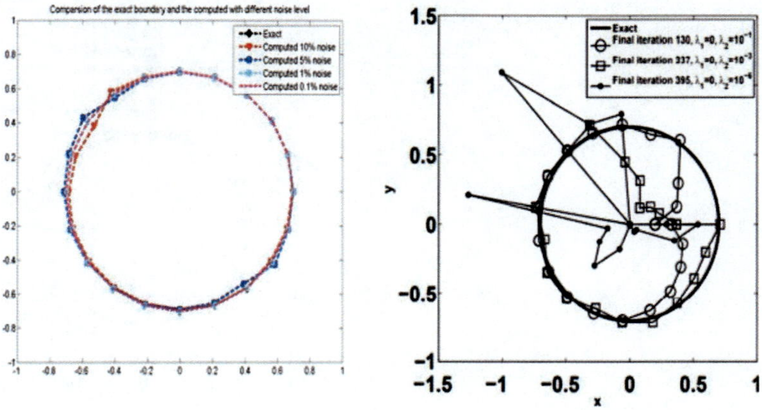

Fig. 3 The reconstructed boundary for Example 1, (Fig. 3a) Decoupled Cauchy-Newton algorithm and (Fig. 3b) Minimisation of the functional

the minimisation process runs until the final convergence is reached, i.e. at 2878 iterations, is unstable.

This instability persists even when a discrepancy-type criterion is used, namely terminating the minimisation process at the first iteration for which the objective function (6) becomes less than the error quantity on g.

This illustrated by solution produced at the iteration number 1267 in Fig. 2b. Whereas, we observe in Fig. 2a that the solution produced by our decoupled Cauchy-Newton algorithm is quite stable.

Comparing Figs. 3 and 4 with Fig. 2 shows that the results obtained by the minimization when $\lambda_1 \in \{10^{-6}, 10^{-3}, 10^{-1}\}$, $\lambda_2 = 0$, become slightly better than the results in Fig. 3b although they still look unstable. Much improved stable results can be seen in Fig. 4b if one regularises with $\lambda_1 = 0$, $\lambda_2 \in \{10^{-6}, 10^{-3}, 10^{-1}\}$. But the results obtained with our method remain better even when these are obtained with a noise level of 10%, see Figs. 3a and 4b.

Example 3 In this example, we consider the more complicated bean shaped inner boundary ∂D given by the radial parameterization

$$r(\theta) = \frac{0.5 + 0.4 * cos(\theta) + 0.1 sin(2\theta)}{1 + 0.7 cos(\theta)}, \quad \theta \in [0, 2\pi] \tag{23}$$

within the unit circle $\hat{A}\Omega = B(0, 1)$. The Dirichlet data (3) on $\partial\Omega$ was taken to be the same as in Example 1. In this case, since no analytical solution is available, the Neumann data (5) is simulated numerically by solving, using the polynomial expansion, the direct Dirichlet problem is considered with $F(x, y) = 0$ and $k = \sqrt{2}$ in Eq. (2), $u_0(r(\theta), \theta) = 0$ on ∂D and $u(1, \theta) = f(\theta) = e^{cos(\theta)+sin(\theta)}$ on $\partial\Omega$,, when ∂D is given by (23).

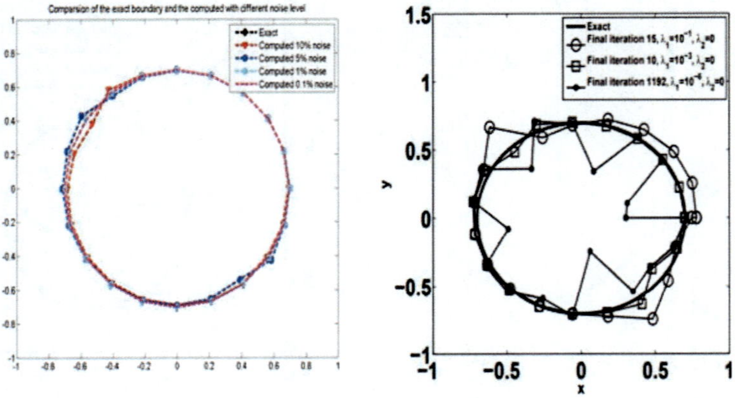

Fig. 4 The reconstructed boundary for Example 1: (Fig. 4a) Decoupled Cauchy-Newton algorithm and (Fig. 4b) Minimisation of the functional with $\lambda_1 = 0$, $\lambda_2 \in \{10^{-6}, 10^{-3}, 10^{-1}\}$

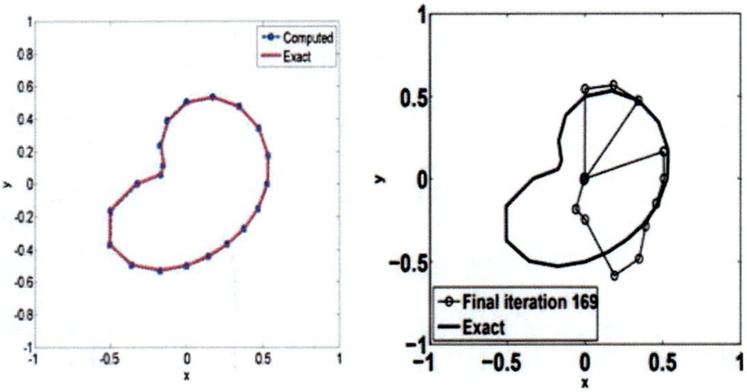

Fig. 5 The reconstructed boundary for Example 3 (Fig. 5a) Decoupled Cauchy-Newton algorithm and (Fig. 5b) Minimisation of the functional

In Fig. 5, we present the reconstructed and the exact bean-shaped inner boundary defined by the radius $r(\theta)$ in (23) obtained with no noise and no regularisation. Our method reconstructs a very accurate inner boundary (see Fig. 5a) when it can be seen From Fig. 5b that the reconstructed shape using the minimisation process fails to determine the exact shape of the inner boundary.

The results displayed in the previous figures and other results not presented here show the superiority of our decoupled Cauchy-Newton algorithm over the minimization method presented in [54]. We will now present other results to verify that the method allows to reconstruct different types of interior boundaries and that this reconstruction is independent of the initial iteration.

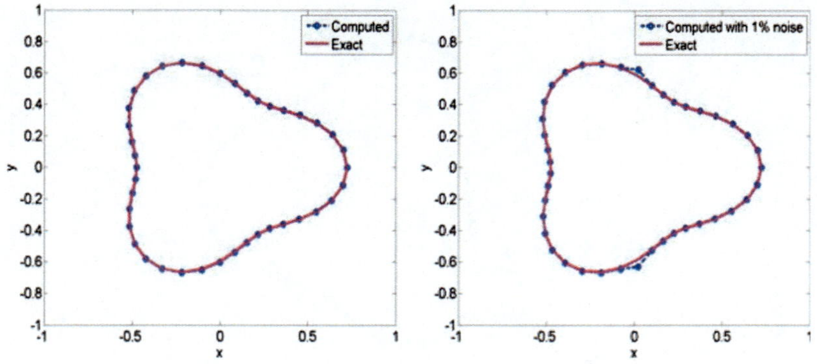

Fig. 6 Example 4 Reconstructed boundaries, (Fig. 6a) no noise, (Fig. 6b) with 1% noise

Example 4 We consider reconstructing the inner boundary ∂D which is a pear-shaped defined by the following parametrization

$$r_D(\theta) = 0.6 + 0.125 * cos(3\theta), \quad \theta \in [0, 2\pi]. \tag{24}$$

The domain Ω and the Dirichlet data (3) on the boundary $\partial\Omega$ are given as in Example 1. The Neumann data (5) on $\partial\Omega$ is simulated numerically by solving the direct Dirichlet problem using polynomial expansion, considered with $F(x, y) = 0$ and $k = 1$ in Eq. (2), $u_0(r(\theta), \theta) = 0$ on ∂D defined by (24). The normal derivative $\partial_\nu u(1, \theta)$ on $\partial\Omega$ is computed using (11).

In Fig. 6 we have presented the internal boundaries for Example 4. The given data for the result in Fig. 6a is without noise and it was obtained without any regularization. In Fig. 6b, the boundary is obtained with 1% noise in the Dirichlet data, We did not add noise in the Neumann condition since it is simulated and therefore it is already noisy. This result was obtained without any regularization or preconditioning. We observe that the noiseless reconstructed curve coincides perfectly with the exact curve. We also observe from Fig. 6b, that this solution remains stable with a noise of 1% these results are obtained with initial iterations for Newton's method which are taken on the circle as in Example 1.

It is clear from Fig. 7 that the reconstructed boundary computed with a bean shaped as the initial iterate is obtained with the same precision as when the initial iterate is a circle. This means that no influence of initial guess is to be retained.

Example 5 In this last example, we consider the problem of reconstructing an internal boundary given by the Eq. (25)

$$r_D(\theta) = 2 * (exp(sin(\theta)) * (sin(2 * \theta))^2 + exp(cos(2 * \theta)) * (cos(2 * \theta))^2) \tag{25}$$

We still keep the Dirichlet condition on the boundary $\partial\Omega$ as in the other examples. The Neumann condition on this boundary is simulated as a normal derivative of the

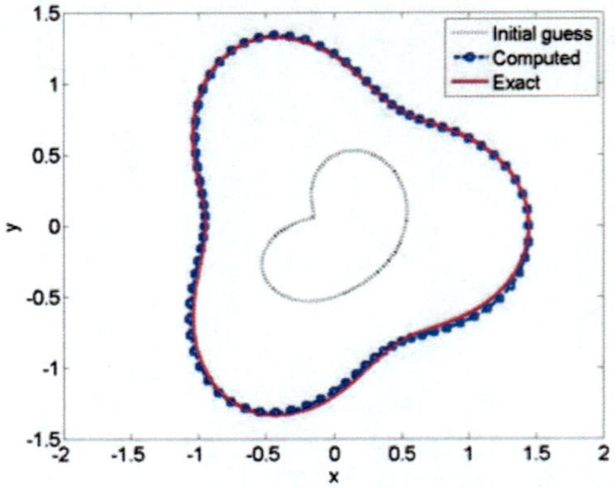

Fig. 7 Example 4 Reconstructed boundaries with initial guess given by (23)

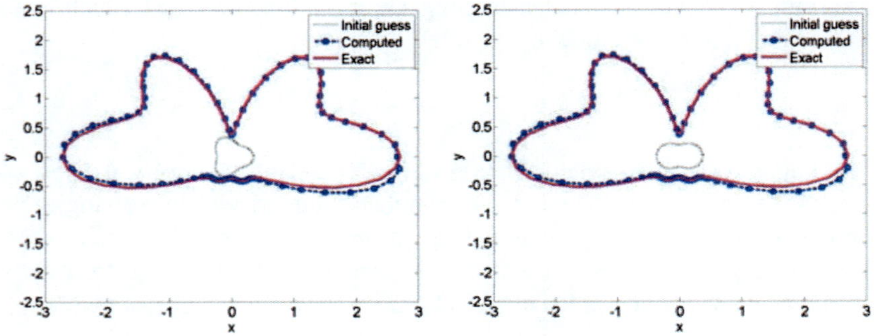

Fig. 8 Example 5 Reconstructed boundaries, (Fig. 8a) with initial guess given by (24), (Fig. 8b) with initial guess given by (26)

solution of the direct problem where $u = 0$ on the boundary of D given by Eq. (25). The other data are taken as in the previous examples.

The result obtained with the initial iterate defined by (24) is presented in Fig. 8a. The reconstructed boundary from an initial guess defined as

$$r = ((3/4) * \sqrt{((cos(\theta))^2 + 0.25 * (sin(\theta))^2)})) \tag{26}$$

is shown in Fig. 8b.

Again we observe, from Fig. 8, that the reconstructed boundaries are very close to the exact boundary. We do not observe from these two figures any difference between the two reconstructed boundaries, which means that the initial guess has no impact on the quality of the reconstructed solution. We deduce that the method is insensitive to the choice of the initial guess.

6 Conclusion

An inverse Cauchy problem for modified Helmholtz equations is considered in a (regular or irregular) domain. The unknown data on a part of the boundary are recovered from the under-specified boundary conditions given on the Cauchy data on the accessible part. The inverse Cauchy problem is transformed to solve a direct problem, using polynomial expansion. The unknown boundary is computed using Newton method. Different cases of regular and irregular boundary are considered. A comparison with the method proposed by [54] the proposed method admit to attend the exact solution with an importantly less cost and with a high accuracy. A noise applied to confirm the stability of the proposed method.

The proposed method has several advantages over the one presented in [54]. First, it does not require any knowledge of the defect a priori. Second, it is well-suited for simply-connected domains with smooth boundaries. Third, the method is insensitive to the choice of the initial guess. In conclusion, we have presented a novel method for solving the inverse problem of determining an unknown defect in a bounded domain from boundary data. The method involves solving a Cauchy problem for the temperature followed by a series of nonlinear scalar equations to determine the boundary of the defect. The proposed method is accurate and efficient and can be applied to a wide range of practical problems in engineering and science.

References

1. C.H. Huang, W.C. Chen, A three-dimensional inverse forced convection problem in estimating surface heat flux by conjugate gradient method. Int. J. Heat Mass Transf. **43**(17), 3171–3181 (2000)
2. A. Nachaoui, An improved implementation of an iterative method in boundary identification problems. Numer. Algorithms **33**(1–4), 381–398 (2003)
3. T. Regińska, K. Regiński, Approximate solution of a Cauchy problem for the Helmholtz equation. Inverse Probl. **22**(3), 975–989 (2006)
4. A. Arsenashvili, A. Nachaoui, T. Tadumadze, On approximate solution of an inverse problem for linear delay differential equations. Bull. Georgian Natl. Acad. Sci. (N.S.) **2**(2), 24–28 (2008)
5. C.L. Fu, X.L. Feng, Z. Qian, The Fourier regularization for solving the Cauchy problem for the Helmholtz equation. Appl. Numer. Math. **59**(10), 2625–2640 (2009)
6. A. Chakib, A. Nachaoui, A. Zeghal, A shape optimization approach for an inverse heat source problem. Int. J. Nonlinear Sci. **12**(1), 78–84 (2012)
7. A. Boulkhemair, A. Nachaoui, A. Chakib, A shape optimization approach for a class of free boundary problems of Bernoulli type. Appl. Math. **58**(2), 205–221 (2013)
8. M.M. Lavrentiev, *Some Improperly Posed Problems of Mathematical Physics* (Springer, 2013)
9. A. Chakib, M. Johri, A. Nachaoui, M. Nachaoui, On a numerical approximation of a highly nonlinear parabolic inverse problem in hydrology. An. Univ. Craiova Ser. Mat. Inform. **42**, 192–201 (2015)
10. V. Isakov, Inverse problems for partial differential equations. *Applied Mathematical Sciences*, vol. 127 (Springer, Cham, 2017)
11. C.S. Liu, F. Wang, A meshless method for solving the nonlinear inverse Cauchy problem of elliptic type equation in a doubly-connected domain. Comput. Math. Appl. **76**, 1831–1852 (2018)

12. K.A. Berdawood, A. Nachaoui, R. Saeed, M. Nachaoui, F. Aboud, An alternating procedure with dynamic relaxation for Cauchy problems governed by the modified Helmholtz equation. Adv. Math. Models Appl. **5**(1), 131–139 (2020)
13. A. Chakib, A. Hadri, A. Laghrib, On a multiscale analysis of an inverse problem of nonlinear transfer law identification in periodic microstructure. Nonlinear Anal. Real World Appl. **51**, 102985 (2020)
14. F. Aboud, A. Nachaoui, M. Nachaoui, On the approximation of a Cauchy problem in a non-homogeneous medium. J. Phys.: Conf. Ser. **1743**(1), 012003 (2021)
15. F. Wang, Y. Gu, W. Qu, C. Zhang, Localized boundary knot method and its application to large-scale acoustic problems. Comput. Methods Appl. Mech. Eng. **361**, 112729 (2020)
16. A. Nachaoui, M. Nachaoui, A. Chakib, M. Hilal, Some novel numerical techniques for an inverse Cauchy problem. J. Comput. Appl. Math. **381**(113030)
17. M. Nachaoui, A. Nachaoui, T. Tadumadze, On the numerical approximation of some inverse problems governed by nonlinear delay differential equation. RAIRO Oper. Res. **56**, 1553–1569 (2022)
18. L. Afraites, C. Masnaoui, M. Nachaoui, Shape optimization method for an inverse geometric source problem and stability at critical shape. Discrete Contin. Dyn. Syst. Ser. S **15**(1), 1–21 (2022)
19. H. Ouaissa, A. Chakib, A. Nachaoui, M. Nachaoui, On numerical approaches for solving an inverse Cauchy stokes problem. Appl. Math. Optim. **85**(1). http://dx.doi.org/ttps://doi.org/10.1007/s00245-022-09833-8
20. A. Ellabib, A. Nachaoui, A. Ousaadane, Convergence study and regularizing property of a modified robin-robin method for the Cauchy problem in linear elasticity. Inverse Probl. **38**, 075007 (2022)
21. A. Nachaoui, Cauchy's problem for the modified biharmonic equation: ill-posedness and iterative regularizing methods, in *New Trends of Mathematical Inverse Problems and Applications*, Springer Proceedings in Mathematics & Statistics (Springer, Cham, 2023)
22. M. Jourhmane, A. Nachaoui, A relaxation algorithm for solving a Cauchy problem, in *Proceedings of the Second International Conferences on Inverse Problems in Engineering* (Engineering Foundation, 1996), pp. 151–158
23. M. Jourhmane, A. Nachaoui, An alternating method for an inverse Cauchy problem. Numer. Algorithms **21**(1), 247–260 (1999)
24. M. Jourhmane, A. Nachaoui, Convergence of an alternating method to solve the Cauchy problem for Poisson's equation. Appl. Anal. **81**(5), 1065–1083 (2002)
25. V.A. Kozlov, V.G. Maz'ya, A.V. Fomin, An iterative method for solving the Cauchy problem for elliptic equations. Zh. Vychisl. Mat. I Mat. Fiz. **31**, 64–74 (1991)
26. A. Nachaoui, F. Aboud, M. Nachaoui, Acceleration of the KMF algorithm convergence to solve the Cauchy problem for Poisson's equation, in *Mathematical Control and Numerical Applications*, Springer Proceedings in Mathematics & Statistics, vol. 372, ed. by A. Nachaoui, A. Hakim, A. Laghrib (Springer, Cham, 2021), pp. 43–57
27. K. Berdawood, A. Nachaoui, M. Nachaoui, An accelerated alternating iterative algorithm for data completion problems connected with Helmholtz equation. Stat. Optim. Inf. Comput. **11**(1), 2–21 (2023). https://doi.org/10.1002/num.22793
28. M. Essaouini, A. Nachaoui, S. El Hajji, Numerical method for solving a class of nonlinear elliptic inverse problems. J. Comput. Appl. Math. **162**(1), 165–181 (2004)
29. M. Essaouini, A. Nachaoui, S. El Hajji, Reconstruction of boundary data for a class of nonlinear inverse problems. J. Inverse Ill-Posed Probl. **12**(4), 369–385 (2004)
30. A. Ellabib, A. Nachaoui, An iterative approach to the solution of an inverse problem in linear elasticity. Math. Comput. Simul. **77**, 189–201 (2008)
31. L. Marin, B.T. Johansson, A relaxation method of an alternating iterative algorithm for the Cauchy problem in linear isotropic elasticity. Comput. Methods Appl. Mech. Eng. **199**(49–52), 3179–3196 (2010)
32. A. Ellabib, A. Nachaoui, A. Ousaadane, Mathematical analysis and simulation of fixed point formulation of Cauchy problem in linear elasticity. Math. Comput. Simul. **187**, 231–247 (2021)

33. A. Nachaoui, Numerical linear algebra for reconstruction inverse problems. J. Comput. Appl. Math. **162**, 147–164 (2004)
34. A. Chakib, A. Nachaoui, M. Nachaoui, H. Ouaissa, On a fixed point study of an inverse problem governed by stokes equation. Inverse Probl. **35**, 015008 (2019)
35. A. Bergam, A. Chakib, A. Nachaoui, M. Nachaoui, Adaptive mesh techniques based on a posteriori error estimates for an inverse Cauchy problem. Appl. Math. Comput. **346**, 865–878 (2019)
36. F. Berntsson, V.A. Kozlov, L. Mpinganzima, B.O. Turesson, Iterative Tikhonov regularization for the Cauchy problem for the Helmholtz equation. Comput. Math. Appl. **73**, 163–172 (2017)
37. A. Chakib, A. Nachaoui, Convergence analysis for finite element approximation to an inverse Cauchy problem. Inverse Probl. **22**(4), 1191–1206 (2006)
38. B. Mukanova, Numerical reconstruction of unknown boundary data in the Cauchy problem for Laplace's equation. Inverse Probl. Sci. Eng. **21**, 1255–1267 (2013)
39. S.I. Kabanikhin, *Inverse and Ill-posed Problems* (Walter de Gruyter GmbH & Co. KG, Berlin, 2012)
40. A.L. Qian, X.T. Xiong, Y.-J. Wu, On a quasi-reversibility regularization method for a Cauchy problem of the Helmholtz equation. J. Comput. Appl. Math. **233**(8), 1969–1979 (2010)
41. R. Shi, T. Wei, H.H. Qin, A fourth-order modified method for the Cauchy problem of the modified Helmholtz equation. Numer. Math. Theory Methods Appl. **2**(3), 326–340 (2009)
42. A. Nachaoui, H.W. Salih, An analytical solution for the nonlinear inverse Cauchy problem. Adv. Math. Models Appl. **6**(3), 191–206 (2021)
43. G.M.M. Reddy, P. Nanda, M. Vynnycky, J.A. Cuminato, An adaptive boundary algorithm for the reconstruction of boundary and initial data using the method of fundamental solutions for the inverse Cauchy-Stefan problem. Comput. Appl. Math. **40**(3)
44. S. Yarmukhamedov, I. Yarmukhamedov, Cauchy problem for the Helmholtz equation, in *Ill-posed and Non-classical Problems of Mathematical Physics and Analysis*, Inverse Ill-posed Probl. Ser., (VSP, Utrecht, 2003), pp. 143–172
45. F. Berntsson, V.A. Kozlov, L. Mpinganzima, B.O. Turesson, An alternating iterative procedure for the Cauchy problem for the Helmholtz equation. Inverse Probl. Sci. Eng. **22**, 45–62 (2014)
46. Q. Hua, Y. Gu, W. Qu, W. Chen, C. Zhang, A meshless generalized finite difference method for inverse Cauchy problems associated with three-dimensional inhomogeneous Helmholtz-type equations. Eng. Anal. Bound. Elem. **82**, 162–171 (2017)
47. Z. Qian, X. Feng, A fractional Tikhonov method for solving a Cauchy problem of Helmholtz equation. Appl. Anal. **96**, 1656–1668 (2017)
48. M. Nachaoui, A. Laghrib, M. Hakim, A new space-variant optimization approach for image segmentation, in *Mathematical Control and Numerical Applications*, ed. by A. Nachaoui, A. Hakim, A. Laghrib (Springer, Cham, 2021), pp. 87–97
49. F. Yang, P. Zhang, X.X. Li, The truncation method for the Cauchy problem of the inhomogeneous Helmholtz equation. Appl. Anal. **98**, 991–1004 (2019)
50. F. Wang, Z. Chen, P.-W. Li, C.-M. Fan, Localized singular boundary method for solving Laplace and Helmholtz equations in arbitrary 2D domains. Eng. Anal. Bound. Elem. **129**, 82–92 (2021)
51. K.A. Berdawood, A. Nachaoui, M. Nachaoui, F. Aboud, An effective relaxed alternating procedure for Cauchy problem connected with Helmholtz equation. Numer. Methods Partial Differ. Equ. 1–27 (2021)
52. K. Berdawood, A. Nachaoui, R. Saeed, M. Nachaoui, F. Aboud, An efficient D-N alternating algorithm for solving an inverse problem for Helmholtz equation. Discrete Contin. Dyn. Syst. Ser. S
53. F. Aboud, I.T. Jameel, A.F. Hasan, B.K. Mostafa, A. Nachaoui, Polynomial approximation of an inverse Cauchy problem for Helmholtz type equations. Adv. Math. Models Appl. **7**(3), 306–322 (2022)
54. B. Bin-Mohsin, The method of fundamental solutions for Helmholtz-type problems, Ph.D. thesis at the University of Leeds
55. S.M. Rasheed, A. Nachaoui, M.F. Hama, A.K. Jabbar, Regularized and preconditioned conjugate gradient like-methods methods for polynomial approximation of an inverse Cauchy problem. Adv. Math. Models Appl. **6**(2), 89–105 (2021)

56. C.S. Liu, C.L. Kuo, A multiple-scale pascal polynomial triangle solving elliptic equations and inverse Cauchy problems. Eng. Anal. Bound. Elem. **62**, 35–43 (2016)
57. K. Boumzough, A. Azzouzi, A. Bouidi, The incomplete LU preconditioner using both CSR and CSC formats. Adv. Math. Models Appl. **7**(2), 156–167 (2022)
58. S.M. Rashid, A. Nachaoui, A haar wavelets-based direct reconstruction method for the Cauchy problem of the Poisson equation. Discrete Contin. Dyn. Syst. Ser. S

Image Restoration Using a Coupled Reaction-Diffusion Equations

Abdelmajid El Hakoume, Ziad Zaabouli, Amine Laghrib, and Lekbir Afaites

Abstract This work addresses In the present paper, we propose a a novel denoising model based on nonlinear-reaction-diffusion system. This model is driven by using the image decomposition strategy of H^{-1} norm given by Guo et al. [1]. This choice is motivated by the success of the decomposition strategy in preserving and illustrating the small features in the textured image. Our results include a rigorous discretization of the proposed model, and numerical results are provided with numerous comparisons to show the performance of the proposed approach.

Keywords Image denoising · Reaction diffusion · PDE

1 Introduction

The use of digital imaging devices is widespread in many areas, including remote sensing and facial detection. The acquired image is a degraded version of the latent observation, and the processing used to improve it is influenced by factors like noise corruption and illumination. Particularly, the unidentified latent observation causes noise to be produced throughout the transmission and compression procedures. To eliminate the noise and retrieve the latent observation from the provided damaged image, image denoising techniques are crucial.

Image denoising methods have gained great interest in recent years, it is an essential issue in the field of image processing that aims to recover a clean image from a noisy one. A noisy image task is generally formulated as follows

$$f = X + N \tag{1}$$

where f is the observed noisy image, X is the corresponding clean image for recovery, and N is the noise component.

A. El Hakoume (✉) · Z. Zaabouli · A. Laghrib · L. Afaites
MI FST Béni-Mellal, Université Sultan Moulay Slimane, Beni Mellal, Morocco
e-mail: abdelmajid.elhakoume@gmail.com

A. Laghrib et al. (eds.), *New Trends of Mathematical Inverse Problems and Applications*, Springer Proceedings in Mathematics & Statistics 428,
https://doi.org/10.1007/978-3-031-33069-8_9

In literature, there exists a multitude of denoising methods, among them we name the variational approaches [2–8], which are recognized for their efficacy in image denoising, wavelet-based technique [9], stochastic approach [10], and patch-based technique [11]. The variational methods gave rise to PDE-based denoising models, which may be divided into two types, namely variational and non-variational PDE models.

To solve the inverse problem (1), then, regularization techniques are essential to construct a stable solution, which indicate some type of a priori knowledge about the solution image X, this allows transforming (1) into a minimization problem as

$$\min_X E(X) = F(X, f) + \lambda R(X), \tag{2}$$

where F is the data fidelity, R is a regularization term and $\lambda > 0$ is a positive parameter that ensure the balance between the two terms. In the literature, there exists a range of regularization terms that has gotten a bit closer to tackling the problem (1), presented by variational minimization and PDEs-based approaches. A common manner of all these approaches is to find adequate function spaces of fidelity term F to describe oscillating patterns and suitable regularizing terms R to address specific difficulties, e.g. the staircase effect.

Inspired by all these ideas, Rudin et al. [12] proposed a famous model namely the total variation (TV) minimization, this model is governed by the following minimization

$$\min_u E(X) = \lambda \|X - f\|^2_{L^2(\Omega)} + \int_\Omega |\nabla X|, \tag{3}$$

where Ω is an open and bounded domain of $\mathbb{R}^2$ and $f \in L^2(\Omega)$. The associated time dependent partial differential equation of the Euler-Lagrange equation $E(X)$ is supplied by the following equation

$$\begin{cases} \frac{\partial X}{\partial t} = div\left(\frac{\nabla X}{|\nabla X|}\right) + 2\lambda(f - X) \ \ in \ \ \Omega, \\ \langle \nabla X, n \rangle = 0 \ \ on \ \ \partial\Omega. \end{cases} \tag{4}$$

This earlier model has been the topic of various research (see [13, 14]) and it is a powerful way to maintain edges throughout the restoration process. Thanks to the effectiveness of TV-based diffusion, numerous changes have been examined (see [15, 16]). Though the TV model performs quite well for image denoising and edges guarding, it may also lose minor features, such as textures (see [17]). Meyer [18] proposed a new minimization method to overcome this limitation by incorporating a weaker norm that is more adequate for capturing oscillatory patterns and tiny features in the textured image. After that, the 4th order denoising models are proposed by Osher et al. [17] using the TV minimization combined with the H^{-1} norm which is defined by

$$\| \cdot \|_{H^{-1}(\Omega)} = \int_\Omega |\nabla \Delta^{-1}(\cdot)|^2 dx. \tag{5}$$

In this approach, the energy functional takes the form

$$\min E_{H^{-1}}(X) = \lambda \|X - f\|^2_{H^{-1}(\Omega)} + \int_{\Omega} |\nabla X|, \tag{6}$$

this functional is naturally linked to the PDE provided by:

$$\begin{cases} \frac{\partial X}{\partial t} = \Delta\Big(div\Big(\frac{\nabla X}{|\nabla X|}\Big)\Big) + 2\lambda(f - X) \;\; in \;\; \Omega, \\ \langle \nabla X, n\rangle = 0 \;\; on \;\; \partial\Omega, \end{cases} \tag{7}$$

This model is named the $TV - H^{-1}$ denoising model which is a strongly nonlinear PDE with no evident numerical solution. To overcome this challenge, as described in [19], the cartoon-texture decomposition technique has been introduced. The image decomposition task involves dividing a given image into two components representing the geometrical and textural parts. Indeed, these strategies give more information about the texture that we are attempting to extract or maintain. According to this technique, the model split the image into two parts. The cartoon information of the original image is represented by X, which denotes the smoothed version of the original image f. Additionally, the component $f - X$ represents the texture or noise information. The Euler-Lagrange of the functional energy (6) take the form

$$\Delta^{-1}(f - X) = \frac{1}{2\lambda} div\Big(\frac{\nabla X}{|\nabla X|}\Big), \tag{8}$$

To solve the difficulty of 4th order degrees in (8), the authors reform it into the following equations:

$$\begin{cases} -div\Big(\frac{\nabla X}{|\nabla X|}\Big) + 2\lambda Y = 0 \;\; in \;\; \Omega, \\ -\Delta Y + (f - X) = 0 \;\; in \;\; \Omega, \\ \langle \nabla X, n\rangle = \langle \nabla Y, n\rangle = 0 \;\; on \;\; \partial\Omega, \end{cases} \tag{9}$$

then, the evolution equation associated with the Eq. (9) is given as follows:

$$\begin{cases} \frac{\partial X}{\partial t} = div\Big(\frac{\nabla X}{|\nabla X|}\Big) - 2\lambda Y \;\; in \;\; \Omega \times (0, T), \\ \frac{\partial Y}{\partial t} = \Delta Y - (f - X) \;\; in \;\; \Omega \times (0, T), \\ \langle \nabla X, n\rangle = \langle \nabla Y, n\rangle = 0 \;\; on \;\; \partial\Omega \times (0, T), \\ X(0, x) = f \;\; in \;\; \Omega, \\ Y(0, x) = 0 \;\; in \;\; \Omega. \end{cases} \tag{10}$$

Inspired by the ideas discussed above, and the efficiency of image decomposition strategies, image-denoising models have seen considerable development latterly to overcome the weakness of the model introduced by Osher et al. [17]. The models vary from paper to another, each has its drawbacks and advantages, and the goal is to give a rigorous analysis and a real interpretation of the denoising task. We refer the reader to extensive studies [4, 20–22] which have been showing promising results.

The principal idea described in this current paper consists of a new non-linear reaction-diffusion system, which is able to efficiently compute the cartoon and texture components of an initial noisy image while retaining its quality and lowering its noise. We propose a nonlinear high-order system that combines the Weickert model [23] and the regularized model introduced by Catte et al. [24], that is better in reducing noise levels, with the decomposition proposed by Guo et al. [1] to manage diffusivity in smooth areas. This model corrects the drawbacks (edge preservation and under smoothing) of the diffusion term presented in [21], by taking the best of the Perona-Malik process in radial regions, and the robustness of the Weickert filter effect near sharp edges.

The main work we have done in this paper is as follows, in Sect. 2, we describe the new proposed model. Then, the well-posedness results of the system are stated in Sect. 3. In Sect. 4, we present discretization counterpart of the model. Finally, Sect. 5 gives numerical results and comparative experiments to illustrate the efficiency of our model.

2 The Proposed Reaction-Diffusion System

In this part, we have presented a denoising model of non-variational type based on the model explored in [25] by altering the diffusion term in both equations. In one hand, the diffusion term in texture part (X) is modified to be the regularized term taken by Catte et al. [24] in which we replace the first order derivative in the function h with the fractional order derivative to avoid much better the staircasing effect, on the other hand, the cartoon part (Y) is choosen to be the Weickert filter [23], to make use of both the Perona-Malik model [26] and the Weickert model [23] in both texture part and carton part. Hence our new model is as follow

$$\begin{cases} \dfrac{\partial X}{\partial t} - div(h(\mid \nabla^{\alpha} X_{\sigma} \mid)\nabla X) + \delta Y = 0 \;\; in \;\; \Omega \times (0,T), \\ \dfrac{\partial Y}{\partial t} - div(\mathcal{D}(J_{\rho}(\nabla Y_{\sigma}))\nabla Y + f - X = 0 \quad in \;\; \Omega \times (0,T), \\ \langle \mathcal{D}(J_{\rho}(\nabla Y_{\sigma}))\nabla Y, n\rangle = \langle h(\nabla X_{\sigma})\nabla X, n\rangle = 0 \;\; on \;\; \partial\Omega \times (0,T), \\ X = 0 \;\; on \;\; \partial\Omega \times (0,T), \\ X(x,0) = f, \quad Y(x,0) = 0 \quad in \;\; \Omega. \end{cases} \tag{11}$$

where $\mathcal{D}$ is an anisotropic diffusion tensor and J_{ρ} is the structure tensor defined by

$$J_{\rho}(\nabla Y_{\sigma}) = K_{\rho} * (\nabla Y_{\sigma} \otimes \nabla Y_{\sigma}), \tag{12}$$

and $*$ is the convolution product and $\otimes$ is the outer product defined such as

$$\nabla Y_\sigma \otimes \nabla Y_\sigma = \nabla Y_\sigma (\nabla Y_\sigma)^t.$$

Here Y_σ is constructed using a convolution of Y with a Gaussian kernel, while K_ρ and K_σ represent two Gaussian convolution kernels such as $K_\tau(x) = \frac{1}{2\pi\tau^2}\exp(-\frac{|x|^2}{2\tau^2})$. The function $\mathcal{D}$ is calculated using the tensor J_ρ eigenvalues and the eigenvectors as follows

$$\mathcal{D} := g_+(\lambda_1, \lambda_2)\theta_1\theta_1^{\mathrm{T}} + g_-(\lambda_1, \lambda_2)\theta_2\theta_2^{\mathrm{T}} = (\theta_1 \;\; \theta_2)\begin{pmatrix} g_+(\lambda_1, \lambda_2) & 0 \\ 0 & g_-(\lambda_1, \lambda_2) \end{pmatrix}(\theta_1 \;\; \theta_2)^t \quad (13)$$

where θ_1, θ_1 are the eigenvectors associated to the eigenvalues λ_1, λ_2 of the tensor structure J_ρ. While the functions g_+ and g_- describe the isotropic or anisotropic behavior of the smoothing on the image regions. For more details about the choice of these functions, we refer the reader to [27].

The idea of the function h is that the smoothing achieved by the equation is conditional, if $| \nabla^\alpha X_\sigma |$ is big, the diffusion will be low and therefore the accurate localization of the edges will be kept. If $| \nabla^\alpha X_\sigma |$ is low, then the diffusion will tend to smooth still more around (x, y). An examples of the function h are given by

$$h(s) = \frac{1}{1+s^2}, \quad h(s) = e^{-s}.$$

The operator ∇^α is the fractional gradient defined in [2], within a constant, such as

$$\nabla^\alpha X_\sigma = \int_\Omega \frac{\big(\nabla X_\sigma(x) - \nabla X_\sigma(y)\big)}{|x-y|^{(\alpha-1)+1}} \frac{x-y}{|x-y|} dy,$$

by setting $s = \alpha - 1$, we have

$$\nabla^\alpha X_\sigma = \int_\Omega \frac{\big(\nabla X_\sigma(x) - \nabla X_\sigma(y)\big)}{|x-y|^{s+1}} \frac{x-y}{|x-y|} dy,$$

Before investigating the well-posedness of the solution for the problem (11), we introduce the following set of hypotheses (H):

- H_1 - The tensors $\mathcal{D}$ is in $C^\infty(\mathbb{R}^{2\times 2}, \mathbb{R}^{2\times 2})$, coercive with the coercivity constant is β, and positive-definite matrix.
- H_2 - The function f is in $L^2(\Omega)$.
- H_3 - The constants ρ, δ, τ, β and σ are positives.
- H_4 - $h : \mathbb{R}^+ \longrightarrow \mathbb{R}^+$ be a smooth decreasing function where $\lim_{s\to+\infty} h(s) = 0$, $h(0) = 1$ and there exists a constant $\mu > 0$ such that $h(|\nabla^\alpha u_\sigma|) \geq \mu$ a.e in $(0, T) \times \Omega$.

3 Existence and Uniqueness

The solution of the suggested denoising model is understood in the sens of the following definition.

Definition 1 A pair of measurable functions (X, Y) is a weak solution of the model (11) if satisfies

$$(X, Y) \in \left(L^2(0, T; H^1(\Omega))\right)^2 \ with \ (\frac{\partial X}{\partial t}, \frac{\partial Y}{\partial t}) \in \left(L^2(0, T; H^1(\Omega)')\right)^2,$$

such that

$$\begin{cases} \langle \frac{\partial X}{\partial t}, \phi \rangle + \int_\Omega h(| \nabla^\alpha X_\sigma |) \nabla X \nabla \phi + \int_\Omega \delta Y \phi = 0, \\ \langle \frac{\partial Y}{\partial t}, \varphi \rangle + \int_\Omega \mathcal{D}(J_\rho(\nabla Y_\sigma)) \nabla Y \nabla \varphi + \int_\Omega f \varphi - \int_\Omega X \phi = 0, \\ \forall \phi, \varphi \in H^1(\Omega). \end{cases} \tag{14}$$

The following lemma summarized a priori estimations of the solution.

Lemma 1 *Under the hypothesis* (H)*, the solution of the problem (14) satisfies the following estimations:*

$$\| X \|_{L^2(0,T;H^1(\Omega))} + \| X \|_{L^\infty(0,T;L^2(\Omega))} + \| \partial_t X \|_{L^2(0,T;H^1(\Omega)')} \leq C,$$

$$\| Y \|_{L^2(0,T;H^1(\Omega))} + \| Y \|_{L^\infty(0,T;L^2(\Omega))} + \| \partial_t Y \|_{L^2(0,T;H^1(\Omega)')} \leq C,$$

with absolute positive constants C.

The following theorem establishes the existence and uniqueness of a solution to the provided coupled system.

Theorem 1 *Under the assumptions* (H) *above, the problem (11) have a unique weak solution* $(X, Y) \in C(0, T; L^2(\Omega)) \cap L^2(0, T; H^1(\Omega))$.

Remark 1 The proof of the the Lemma 1 and the Theorem 1 is similar to the proof given in [22, 27].

4 Discretization

In this part, we present all the necessary discrete operators to approximate the spatial and temporal derivatives that appear in our numerical schemes. To do so, we use a fully discretization of the model using finite difference approach.

Let h represent the spatial grid size and k the time step size, we discretize time and space as:

$$\begin{aligned} x_i &= ih, \ i = 0, 1, 2, ..., N, \\ y_j &= jh, \ j = 0, 1, 2, ..., M, \\ t_n &= nk, \ n = 0, 1, 2..... \end{aligned}$$

Let $X^n_{i,j}$, $Y^n_{i,j}$ and $f_{i,j}$ denote the approximations of $X(t_n, x_i, y_j)$, $Y(t_n, x_i, y_j)$ and $f(x, y)$ respectively, then the discretization of the time derivative is given by:

$$\frac{\partial X(t, x, y)}{\partial t} = \frac{X^{n+1}_{i,j} - X^n_{i,j}}{k} \text{ and } \frac{\partial Y(t, x, y)}{\partial t} = \frac{Y^{n+1}_{i,j} - Y^n_{i,j}}{k}.$$

the discretization of the Perona-Malik term $div(h(\mid \nabla^\alpha X_\sigma \mid)\nabla X)$ can be as follows: we denote by $[A(X^n)]_{i,j}$ the approximation of $div(h(\mid \nabla X_\sigma(t_n, x_i, y_j) \mid))\nabla X(t_n, x_i, y_j))$ defined by

$$[A(X^n)]_{i,j} = h_{i+\frac{1}{2},j}(X^n_{i+1,j} - X^n_{i,j}) - h_{i-\frac{1}{2},j}(X^n_{i+,j} - X^n_{i-1,j}) + h_{i,j+\frac{1}{2}}(X^n_{i,j+1} - X^n_{i,j}) - h_{i,j-\frac{1}{2}}(X^n_{i,j} - X^n_{i,j-1})$$

where

$$h_{i\pm\frac{1}{2},j} = \frac{h_{i\pm1,j} - h_{i,j}}{2}, \quad h_{i,j\pm\frac{1}{2}} = \frac{h_{i,j\pm1} - h_{i,j}}{2}.$$

The components of each $h_{i,j}$ contain the discretization of the fractional derivative figuring in h, for that reason we put the zero Dirichlet boundary condition on X which is figuring in (11). Now, let's treat the diffusion term $\mathrm{div}(\mathcal{D}(J_\rho(\nabla Y_\sigma))\nabla Y)$. In fact, there exist various approaches in the literature that enable to discretize this term. Here, we explore a traditional approach which presented by Weickert in [23]. Let us now denoted

$$\mathcal{D}(J_\rho(\nabla Y_\sigma)) = \begin{pmatrix} a & b \\ b & c \end{pmatrix}, \tag{15}$$

with

$$\begin{aligned} a &= \mu_1 \cos^2 \alpha + \mu_2 \sin^2 \alpha, \\ b &= (\mu_1 - \mu_2) \sin \alpha \cos \alpha, \\ c &= \mu_1 \sin^2 \alpha + \mu_2 \cos^2 \alpha. \end{aligned}$$

Where α verifies the formula

$$\tan(2\alpha) = \frac{2J_{12}}{J_{11} - J_{22}},$$

with J_{12}, J_{11} and J_{22} are the components of the structure tensor J_ρ, that are

$$J_\rho(\nabla v_\sigma) = \begin{pmatrix} J_{11} & J_{12} \\ J_{12} & J_{22} \end{pmatrix}$$

So, the diffusion term $\operatorname{div}(\mathcal{D}(J_\rho(\nabla Y_\sigma))\nabla Y^n)$ can be developed as

$$\begin{aligned}\operatorname{div}(\mathcal{D}(J_\rho(\nabla Y_\sigma))\nabla Y^n) &= \operatorname{div}\left[\begin{pmatrix} a & b \\ b & c \end{pmatrix}\begin{pmatrix} \partial_x Y^n \\ \partial_y Y^n \end{pmatrix}\right] \\ &= \partial_x\left(a\partial_x Y^n\right) + \partial_x\left(b\partial_y Y^n\right) + \partial_y\left(b\partial_x Y^n\right) + \partial_y\left(c\partial_y Y^n\right).\end{aligned}$$

Where

$$\begin{aligned}\partial_x\left(a\partial_x Y^n\right) &= \partial_x a\partial_x Y^n + a\partial_{xx} Y^n,\\ \partial_x\left(b\partial_y Y^n\right) &= \partial_x b\partial_y Y^n + b\partial_{xy} Y^n,\\ \partial_y\left(b\partial_x Y^n\right) &= \partial_y b\partial_x Y^n + b\partial_{yx} Y^n,\\ \partial_y\left(c\partial_y Y^n\right) &= \partial_y c\partial_y Y^n + c\partial_{yy} Y^n.\end{aligned}$$

Let as consider a discrete rectangular image domain with $N \times M$ pixels and denote by $Y^n_{i,j}$, $i = 1, \ldots N$ and $j = 1, \ldots M$ this image. We also denote by $(\partial_x Y^n)_{i,j}$, $(\partial_y Y^n)_{i,j}$, $(\partial_{xx} Y^n)_{i,j}$, $(\partial_{xy} Y^n)_{i,j}$, $(\partial_{yy} Y^n)_{i,j}$ the discrete version respectively of the operators $\partial_x Y^n$, $\partial_y Y^n$, $\partial_{xx} Y^n$, $\partial_{xy} Y^n$, $\partial_{yy} Y^n$, that are

$$\left(\partial_x Y^n\right)_{i,j} = \begin{cases} Y^n_{i+1,j} - Y^n_{i,j} & \text{if } i < N \\ 0 & \text{if } i = N \end{cases}, \quad \left(\partial_y Y^n\right)_{i,j} = \begin{cases} Y^n_{i,j+1} - Y^n_{i,j} & \text{if } j < M \\ 0 & \text{if } j = M \end{cases}. \tag{16}$$

The second-order derivatives operators are approximated as

$$\begin{aligned}(\partial_{xx} Y^n)_{i,j} &= \begin{cases} Y^n_{i+1,j} - 2Y^n_{i,j} + Y^n_{i-1,j} & \text{si } 1 < i < N, \\ Y^n_{i+1,j} - Y^n_{i,j} & \text{si } i = 1, \\ Y^n_{i-1,j} - Y^n_{i,j} & \text{si } i = N \end{cases}\\ \left(\partial_{xy} v^n\right)_{i,j} &= \begin{cases} Y^n_{i,j+1} - Y^n_{i,j} - Y^n_{i-1,j+1} + Y^n_{i-,j} & \text{si } 1 < i < N, 1 < j < M \\ 0 & \text{si } i = 1, \\ 0 & \text{si } i = N \end{cases}\\ \left(\partial_{yx} Y^n\right)_{i,j} &= \begin{cases} Y^n_{i+1,j} - Y^n_{i,j} - Y^n_{i+1,j-1} + Y^n_{i,j-1} & \text{si } 1 < i < N, 1 < j < M \\ 0 & \text{si } j = 1 \\ 0 & \text{si } j = M \end{cases}\\ \left(\partial_{yy} Y^n\right)_{i,j} &= \begin{cases} Y^n_{i,j+1} - 2Y^n_{i,j} + Y^n_{i,j-1} & \text{si } 1 < j < M \\ Y^n_{i,j+1} - Y^n_{i,j} & \text{si } j = 1, \\ Y^n_{i,j-1} - Y^n_{i,j} & \text{si } j = M \end{cases}\end{aligned}$$

Using these notations, the discretization of the diffusion term can be written:

$$\big[\mathrm{div}\,(\mathcal{D}(J_\rho(\nabla Y_\sigma))\nabla Y^n)\big]_{i,j} = (\partial_x a)_{i,j}(\partial_x Y^n)_{i,j} + (a)_{i,j}(\partial_{xx} Y^n)_{i,j} + (\partial_x b)_{i,j}(\partial_y Y^n)_{i,j} \tag{17}$$

$$+(b)_{i,j}(\partial_{xy} Y^n)_{i,j} + (\partial_y b)_{i,j}(\partial_x Y^n)_{i,j} + (b)_{i,j}(\partial_{yx} Y^n)_{i,j} \tag{18}$$

$$+(\partial_y c)_{i,j}(\partial_y Y^n)_{i,j} + (c)_{i,j}(\partial_{yy} Y^n)_{i,j}. \tag{19}$$

In fact, the standard discretization of (17) is equivalent to convolve the discrete image v at any point (i, j) by the Stencil matrix of size 3×3 (Table 1). Moreover, one of the conditions imposed by Weickert [23] on the matrix corresponding to the diffusion term is the positivity of its components, which is crucial for the stability of this schema. If $B(Y^n)$ represents the matrix containing the components of the term div $(\mathcal{D}(J_\rho(\nabla Y_\sigma))\nabla Y^n)$ and the Stencil matrix S defined as:

$$S = \begin{pmatrix} s_{11} & s_{12} & s_{13} \\ s_{21} & s_{22} & s_{23} \\ s_{31} & s_{32} & s_{33} \end{pmatrix}.$$

then, an element of $B(Y^n)$ is calculated by:

$$\begin{aligned}[B(Y^n)]_{i,j} = & Y^n_{i-1,j-1}s_{11} + Y^n_{i-1,j}s_{12} + Y^n_{i-1,j+1}s_{13} \\ & + Y^n_{i,j-1}s_{21} + Y^n_{i,j}s_{22} + Y^n_{i,j+1}s_{23} \\ & + Y^n_{i+1,j-1}s_{31} + Y^n_{i+1,j}s_{32} + Y^n_{i+1,j+1}s_{33}.\end{aligned}$$

Finally, the discrete explicit scheme of the reaction-diffusion system (11) might be expressed as

$$\begin{cases} X^{n+1}_{i,j} = X^n_{i,j} + k(\big[A(X^n)\big]_{i,j} - \delta Y^n_{i,j}), \\ Y^{n+1}_{i,j} = Y^n_{i,j} + k(\big[B(Y^n)\big]_{i,j} - f_{i,j} + X^n_{i,j}), \\ X^0_{i,j} = f_{i,j}, \;\; Y^0_{i,j} = 0. \end{cases} \tag{20}$$

for all $k \geq 0$, $0 \leq j \leq M$ and $0 \leq i \leq N$.

Then, the main algorithm related to the model (11) is stated in Algorithm 1.

5 Numerical Results

Some selective results of image denoising are presented in this section, which will justify the introduced coupled PDE for complex noise removal. Aiming for better introducing the superiorities, our denoising results are compared with those of some state of art denoising methods. In fact we compare with two classical image reconstruction methods, i.e., reaction diffusion PDE (RDPDE) [1], controlled reaction diffusion system (CRDS) [4] and fractional reaction diffusion equation (FRDE)

Table 1 Non negative Stencil discretization

$\frac{\lvert b_{i-1,j+1}\rvert - b_{i-1,j+1}}{4} + \frac{\lvert b_{i,j}\rvert - b_{i,j}}{4}$	$\frac{c_{i,j+1}+c_{i,j}}{2} - \frac{\lvert b_{i,j+1}\rvert + \lvert b_{i,j}\rvert}{2}$	$\frac{\lvert b_{i+1,j+1}\rvert + b_{i+1,j+1}}{4} + \frac{\lvert b_{i,j}\rvert + b_{i,j}}{4}$
$\frac{a_{i-1,j}+a_{i,j}}{2} - \frac{\lvert b_{i-1,j}\rvert + \lvert b_{i,j}\rvert}{2}$	$-\frac{a_{i-1,j}+2a_{i,j}+a_{i+1,j}}{2} - \frac{\lvert b_{i-1,j+1}\rvert - b_{i-1,j+1} + \lvert b_{i+1,j+1}\rvert + b_{i+1,j+1}}{4} - \frac{\lvert b_{i-1,j-1}\rvert + b_{i-1,j-1} + \lvert b_{i+1,j-1}\rvert - b_{i+1,j-1}}{4} + \frac{\lvert b_{i-1,j}\rvert + \lvert b_{i+1,j}\rvert + \lvert b_{i,j-1}\rvert + \lvert b_{i,j+1}\rvert + 2\lvert b_{i,j}\rvert}{2} - \frac{c_{i,j-1}+2c_{i,j}+c_{i,j+1}}{2}$	$\frac{a_{i+1,j}+a_{i,j}}{2} - \frac{\lvert b_{i+1,j}\rvert + \lvert b_{i,j}\rvert}{2}$
$\frac{\lvert b_{i-1,j-1}\rvert + b_{i-1,j-1}}{4} + \frac{\lvert b_{i,j}\rvert + b_{i,j}}{4}$	$\frac{c_{i,j-1}+c_{i,j}}{2} - \frac{\lvert b_{i,j-1}\rvert + \lvert b_{i,j}\rvert}{2}$	$\frac{\lvert b_{i+1,j-1}\rvert - b_{i+1,j-1}}{4} + \frac{\lvert b_{i,j}\rvert - b_{i,j}}{4}$

Input: Given a noisy image f and an initial $Y_0 = 0$, choose ϵ, $\rho,\ \beta,\ \sigma > 0,\ T > 0,\ (k_1, k_2) \in \mathbb{R}^2$.
Output: The restored image X and the carton part Y.

1. **Repeat**
2. Compute X^{n+1} by solving the first equation of the system (20).
3. Compute Y^{n+1} by solving the second equation of the system (20).
4. **Update**
5. **Until** $\dfrac{\| X^{n+1}(T) - X^n(T) \|_{L^2}}{\| X^n(T) \|_{L^2}} \leq \epsilon.$

Algorithm 1: Coupled algorithm for the system (11)

Fig. 1 The used three image for comparison

[20]. We note here that the fidelity term used for the compared method is the same as our approach. The tested algorithms are performed on both grayscale images and color images, and the original data are shown in Fig. 1. In the meanwhile, all experimental results are carried out under the platform of Windows 10 by adopting MATLAB R2015 on a Laptop with an Intel(R) Core(TM) i7 CPU at 3.20 GHz and 16 GB memory.

- We start by a first test, where the *Barbara* image is considered. A speckle noise is added with a parameter $M = 10$ and also a Gaussian noise is added with variance $\sigma^2 = 0.02$. Then, we recover the clean image u using the proposed PDE and we compare it to the obtained one by the other compared methods. The obtained results are depicted in Fig. 2. As seen, the obtained result is more pleasant. We plot in addition the 3D representation of the recovered u in Fig. 3 and the same observation remain true.
- The second test concerns the *Penguin* image, which is contaminated by a speckle noise with a parameter $L = 12$ and also a Gaussian noise is added with variance $\sigma^2 = 0.03$. The obtained results are shown in Fig. 4. The resulting clean image is more pleasant and more coherent than the compared ones. This is confirmed in Fig. 5, where 3D representation of $u(1 : 80, 1 : 60)$ is carried out.
- For the third test, we use the *Chest MRI* image, which is contaminated by a speckle noise with a parameter $L = 15$ and also a Gaussian noise is added with variance $\sigma^2 = 0.04$. The obtained results are shown in Fig. 6. The same previous remarks remain valid for this test too when the 3D surfaces are plotted in Fig. 7.

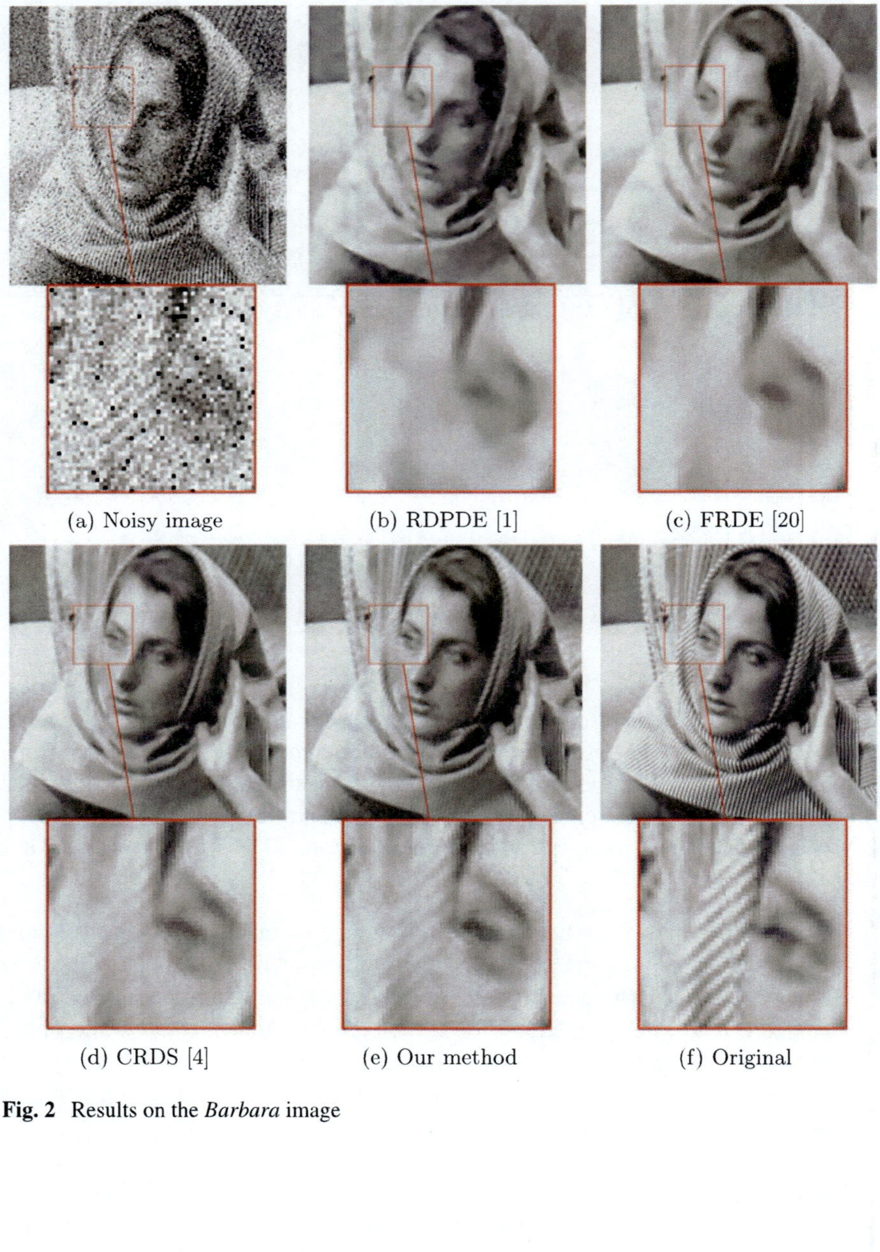

(a) Noisy image (b) RDPDE [1] (c) FRDE [20]

(d) CRDS [4] (e) Our method (f) Original

Fig. 2 Results on the *Barbara* image

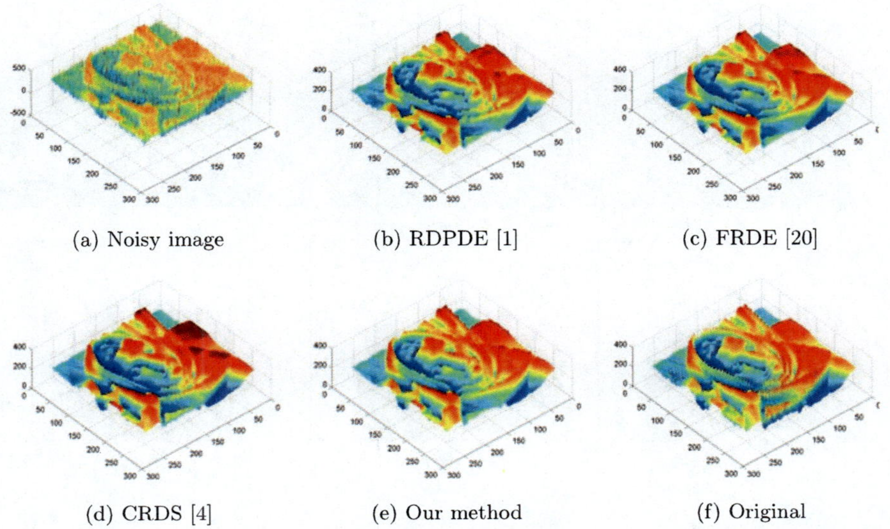

(a) Noisy image (b) RDPDE [1] (c) FRDE [20]

(d) CRDS [4] (e) Our method (f) Original

Fig. 3 The 3D representation of the recovered *Barbara* image u for different denoising approaches

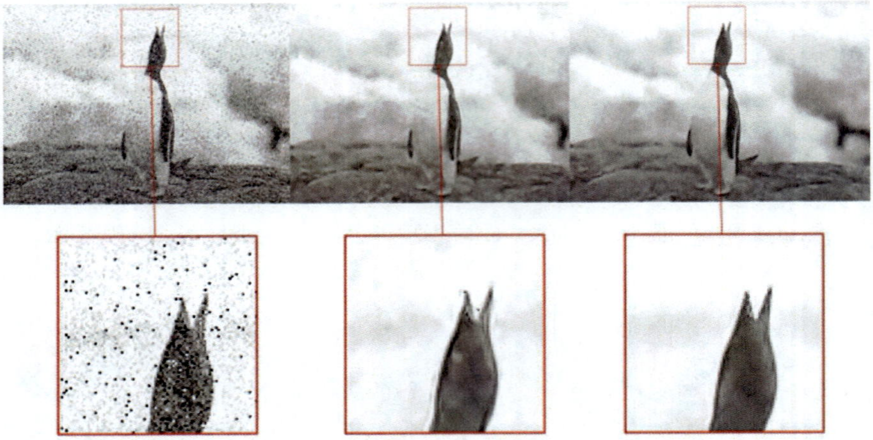

(a) Noisy image (b) RDPDE [1] (c) FRDE [20]

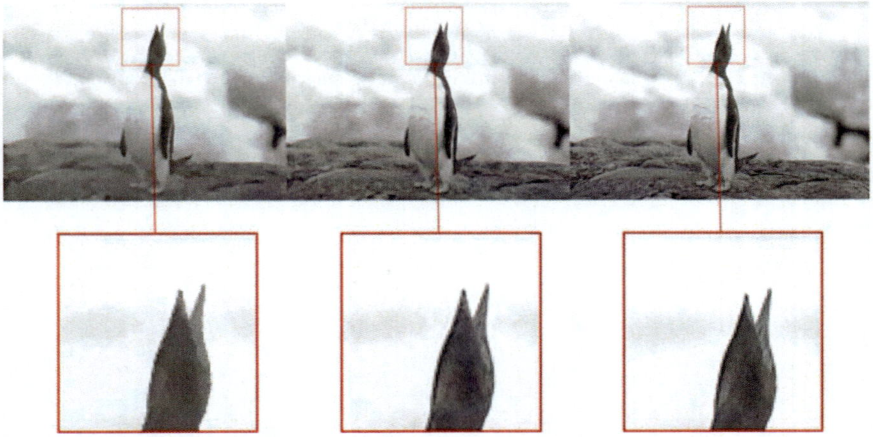

(d) CRDS [4] (e) Our method (f) Original

Fig. 4 Results on the *Penguin* image

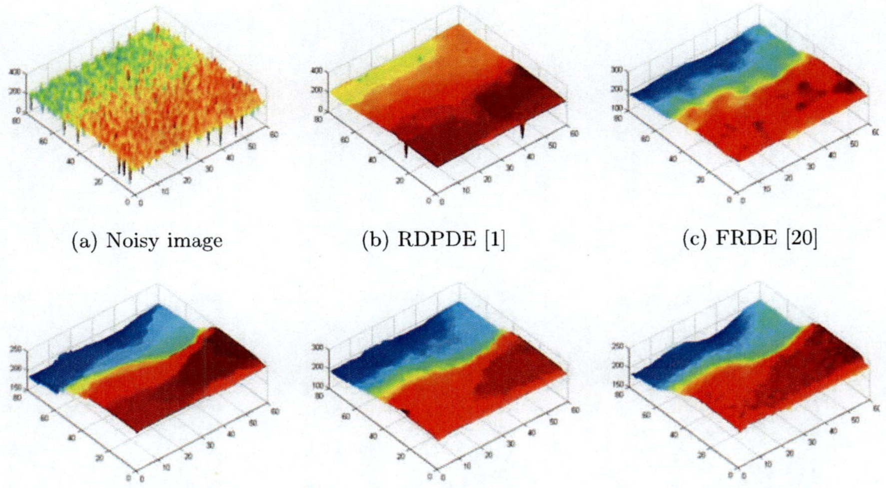

(a) Noisy image (b) RDPDE [1] (c) FRDE [20]

(d) CRDS [4] (e) Our method (f) Original

Fig. 5 The 3D representation of the recovered *Penguin* image $u(1:80, 1:60)$ for different denoising approaches

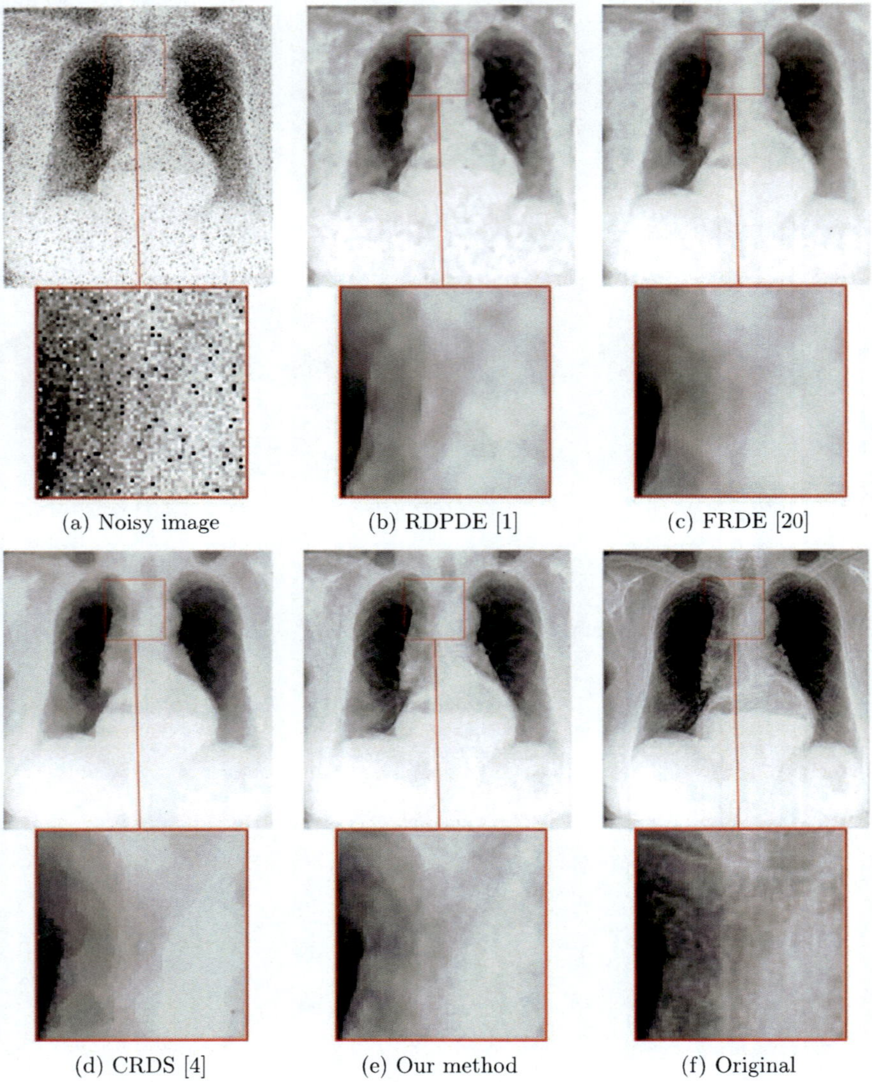

Fig. 6 Results on the *MRI Chest* image

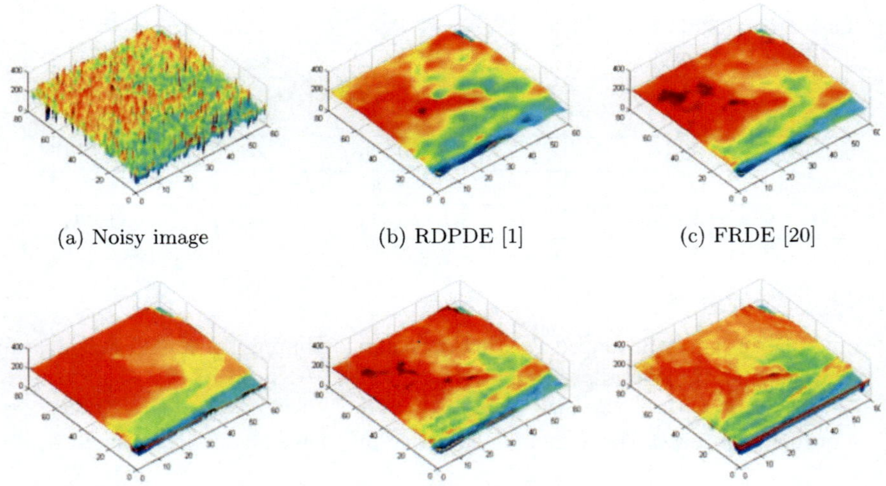

(a) Noisy image (b) RDPDE [1] (c) FRDE [20]

(d) CRDS [4] (e) Our method (f) Original

Fig. 7 The 3D representation of the recovered *MRI Chest* image $u(1:80, 1:60)$ for different denoising approaches

Table 2 The PSNR table

Image	PSNR				
	Noisy	RDPDE [1]	FRDE [20]	CRDS [4]	Proposed approach
Barbara	17.40	27.42	27.88	28.59	**30.28**
Penguin	16.44	26.25	27.02	26.78	**27.71**
MRI Chest	15.66	26.82	27.05	27.54	**28.88**

We follow the qualitative criterion for the three tests by quantitative comparison. All the testing results are evaluated on the average Peak Signal-to-Noise Ratio (PSNR) and Structural-Similarity index (SSIM) of the reconstruction quality. In fact Tables 2 and 3 illustrated the obtained PSNR and SSIM values for the three above tests, respectively. As shown, always the introduced PDE is with the best value, which confirm the visual efficiency.

6 Conclusion

This present paper is about image processing, in particular denoising task. For that we proposed a new reaction-diffusion system based on image-decomposition strategies. The well-posedness result are stated above, further, the simulation results clam that the model could efficiently recover the image features. Future work could focus

Table 3 The SSIM table

Image	PSNR				
	Noisy	RDPDE [1]	FRDE [20]	CRDS [4]	Proposed approach
Barbara	0.563	0.787	0.796	0.806	**0.838**
Penguin	0.440	0.779	0.781	0.784	**0.799**
MRI Chest	0.366	0.672	0.688	0.669	**0.705**

on exploring more advanced PDE models and combining them with deep learning techniques for improved performance.

Conflict of interest The authors declare that they have no conflict of interest.

Acknowledgements This work was entirely funded by the respective institutions of the authors.

References

1. Z. Guo, J. Yin, Q. Liu, On a reaction-diffusion system applied to image decomposition and restoration. Math. Comput. Model. **53**(5–6), 1336–1350 (2011)
2. L. Afraites, A. Hadri, A. Laghrib, A denoising model adapted for impulse and Gaussian noises using a constrained-PDE. Inverse Probl. **36**(2), 025006 (2020)
3. L. Afraites, A. Hadri, A. Laghrib, M. Nachaoui, A high order PDE-constrained optimization for the image denoising problem. Inverse Probl. Sci. Eng. **29**(12), 1821–1863 (2021)
4. A. Hadri, H. Khalfi, A. Laghrib, M. Nachaoui, An improved spatially controlled reaction-diffusion equation with a non-linear second order operator for image super-resolution. Nonlinear Anal. Real World Appl. **62**, 103352 (2021)
5. M. Lysaker, A. Lundervold, X.-C. Tai, Noise removal using fourth-order partial differential equation with applications to medical magnetic resonance images in space and time. IEEE Trans. Image Process. **12**(12), 1579–1590 (2003)
6. A. Lekbir, H. Aissam, L. Amine, N. Mourad, A non-convex denoising model for impulse and Gaussian noise mixture removing using bi-level parameter identification. Inverse Probl. Imaging **16**(4), 827–870 (2022)
7. M. Nachaoui, L. Afraites, A. Laghrib, A regularization by denoising super-resolution method based on genetic algorithms. Signal Process. Image Commun. **99**, 116505 (2021)
8. A. Hadri, L. Afraites, A. Laghrib, M. Nachaoui, A novel image denoising approach based on a non-convex constrained PDE: application to ultrasound images. Signal Image Video Process. **15**, 1057–1064 (2021)
9. S.G. Chang, B. Yu, M. Vetterli, Adaptive wavelet thresholding for image denoising and compression. IEEE Trans. Image Process. **9**(9), 1532–1546 (2000)
10. G.L. Gime farb, *Image Textures and Gibbs Random Fields* (Springer, 1999)
11. A. Buades, B. Coll, J.-M. Morel, A non-local algorithm for image denoising, in *2005 IEEE Computer Society Conference on Computer Vision and Pattern Recognition (CVPR'05)*, vol. 2. IEEE (2005), pp. 60–65
12. L.I. Rudin, S. Osher, E. Fatemi, Nonlinear total variation based noise removal algorithms. Physica D: Nonlinear Phenom. **60**(1–4), 259–268 (1992)

13. R. Acar, C.R. Vogel, Analysis of bounded variation penalty methods for ill-posed problems. Inverse Probl. **10**(6), 1217 (1994)
14. A. Chambolle, P.-L. Lions, Image recovery via total variation minimization and related problems. Numerische Mathematik **76**(2), 167–188 (1997)
15. A. Hadri, A. Laghrib, H. Oummi, An optimal variable exponent model for magnetic resonance images denoising. Pattern Recognit. Lett. **151**, 302–309 (2021)
16. L.A. Vese, S.J. Osher, Modeling textures with total variation minimization and oscillating patterns in image processing. J. Sci. Comput. **19**(1), 553–572 (2003)
17. S. Osher, A. Solé, L. Vese, Image decomposition and restoration using total variation minimization and the h. Multiscale Model. Simul. **1**(3), 349–370 (2003)
18. Y. Meyer, *Oscillating Patterns in Image Processing and Nonlinear Evolution Equations: The Fifteenth Dean Jacqueline B. Lewis Memorial Lectures*, vol. 22, (American Mathematical Society, 2001)
19. Z. Guo, Q. Liu, J. Sun, B. Wu, Reaction-diffusion systems with p (x)-growth for image denoising. Nonlinear Anal. Real World Appl. **12**(5), 2904–2918 (2011)
20. A. Atlas, M. Bendahmane, F. Karami, D. Meskine, O. Oubbih, A nonlinear fractional reaction-diffusion system applied to image denoising and decomposition. Discrete Contin. Dyn. Syst. B **26**(9), 4963 (2021)
21. A. Halim, B.R. Kumar, A TV- L2- H- 1 PDE model for effective denoising. Comput. Math. Appl. **80**(10), 2176–2193 (2020)
22. A. El Hakoume, L. Afraites, A. Laghrib, Well-posedness and simulation results of a coupled denoising PDE. Nonlinear Anal. Real World Appl. **65**, 103499 (2022)
23. J. Weickert et al., *Anisotropic Diffusion in Image Processing*, vol. 1 (Teubner, Stuttgart, 1998)
24. F. Catté, P.-L. Lions, J.-M. Morel, T. Coll, Image selective smoothing and edge detection by nonlinear diffusion. SIAM J. Numer. Anal. **29**(1), 182–193 (1992)
25. A. El Hakoume, L. Afraites, A. Laghrib, An improved coupled PDE system applied to the inverse image denoising problem. Electron. Res. Arch. **30**(7), 2618–2642 (2022)
26. P. Perona, J. Malik, Scale-space and edge detection using anisotropic diffusion. IEEE Trans. Pattern Anal. Mach. Intell. **12**(7), 629–639 (1990)
27. I. El Mourabit, M. El Rhabi, A. Hakim, A. Laghrib, E. Moreau, A new denoising model for multi-frame super-resolution image reconstruction. Signal Process. **132**, 51–65 (2017)

A Novel Identification Scheme of an Inverse Source Problem Based on Hilbert Reproducing Kernels

François Jauberteau, Mourad Nachaoui, and Sara Zaroual

Abstract This work deals with an inverse problem of term source identification. An optimal control formulation is proposed. Thus, an existence result of an optimal solution is established. To solve the obtained optimization problem an approach based on the reproducing kernel Hilbert spaces is proposed. The proposed optimization method based on the gradient descent algorithm is examined through some numerical experiments. The effectiveness of the proposed approach is proven by the presented results.

Keywords Inverse problem · Source term identification · Hilbert spaces with reproducing kernels · Control optimal formulation · Gradient method

1 Introduction

Over the years, inverse problems have undergone a great evolution in several fields of science. In medicine, we cite the study of electroencephalography (EEG) [12] or also the restoration of the activities of the human heart from measurements of potential on the body [39]. In the environment, various works have dealt with the identification of sources of pollution in the atmosphere and in surface waters [13]. Another example is the identification of sources in bioluminescence tomography [1, 4, 5, 28] and the identification of dislocations in materials [7]. Inverse problems can also occur in chemistry to identify chemical reaction constants, in image processing to restore blurred images, and even in the biomedical field to locate faulty

F. Jauberteau
Laboratoire de Mathématiques Jean Leray UMR6629 CNRS/Université de Nantes 2 rue de la Houssinière, BP92208 44322 Nantes, France
e-mail: Francois.Jauberteau@univ-nantes.fr

M. Nachaoui (✉) · S. Zaroual
Equipe de Mathématiques et Interactions, FST Béni-Mellal, Université Sultan Moulay Slimane, Béni-Mellal, Maroc
e-mail: nachaoui@gmail.com

A. Laghrib et al. (eds.), *New Trends of Mathematical Inverse Problems and Applications*, Springer Proceedings in Mathematics & Statistics 428,
https://doi.org/10.1007/978-3-031-33069-8_10

genes and cancer cells. In the energy field, the inverse problem is used to identify underground pockets of oil. Finally, in the fields of freshwater underground flows and saline intrusion, many works have used inverse problems to identify parameters, see for example [2, 3, 8, 17, 23–27, 36].

In this work, we study an inverse source identification problem, which consists in identifying the source term from a measurement on a part of the boundary. We then seek to determine both the second member and the solution of the partial differential equation. Different approaches are used to solve this kind of problem [16, 18], in this work we propose an approach based on Hilbert spaces with reproducing kernels [29].

During the last years, the use of reproducing kernels has known a great interest in solving nonlinear physical problems and engineering problems [34]. The reproducing kernel has been successfully applied to wavelet transforms [37], stochastic processes [22], signal processing [31], machine learning [19–21], ill-posed Cauchy problems for elliptic equations [35], inverse problems [15, 38], etc. Many papers indicate that the kernel method reproducing [6, 9, 32] has exceptional advantages and can handle nonlinear and ill-posed problems. Numerical solutions can also be obtained by this method for practical problems that could not be solved efficiently before [11].

In order to theoretically and numerically study this problem, we reformulate it into an optimal control problem [40], then we study the existence of the optimal solution. Then, we compute the gradient of the cost function by introducing the sensitivity problem and the adjoint problem [10]. Then, we introduce the reproducing kernel space and we give a representation of the optimal solution in the reproducing kernel space. Concerning the approximation of the state problem, we use the finite element method for the discretization of the direct problem and the gradient method for the minimization.

The chapter is organized as follows: After an introduction, Sect. 2 describes the considered inverse problem and its formulation into an optimal control problem. Section 3 presents the study of existence of an optimal solution of the considered optimal control problem. The numerical scheme of the gradient of the cost functional by introducing the sensitivity problem and the adjoint problem is presented in Sect. 4. The numerical experiments are presented in Sect. 5.

2 Setting Problem and Formulation

Our study will focus on the problem of identifying the source of an elliptic PDE based on Dirichlet boundary data. Concretely, based on the following Dirichlet condition

$$\left\{ u = \varphi \ sur \ \Gamma_0, \right. \tag{1}$$

we seek to determine the source term f and the potential u solution of the following problem

$$\begin{cases} -\Delta u + pu = f \ dans \ \Omega, \\ \partial_n u = g \ sur \quad \Gamma. \end{cases} \tag{2}$$

Where Γ_0 is a non empty part of the boundary $\Gamma := \partial\Omega$ of the bounded domain $\Omega \in \mathbb{R}^d$ ($d = 2$ or 3) and $\varphi \in L^2(\Gamma_0)$ is the boundary data, p is a continuous function on $\overline{\Omega}$ that verify $0 < c_0 < p$, n denotes the outwards pointing unit normal vector of the boundary Ω, and $g \in L^2(\Gamma)$ is a given function.

This problem, and variations on it, can be found in a wide range of applications. As an illustration, consider crack detection [41], and the inverse ECG problem [30]. Several techniques for computing trustworthy results have been devised, despite the fact that the majority of source identification problems for elliptic PDEs are poorly posed. As an example, [4, 5, 14] and references therein assume a priori that f is made of a finite number of pointwise sources or sources with compact support inside a limited number of finite subdomains. Such methods give rise to complex mathematical problems, although in many instances, explicit regularization or optimization can be avoided.

In order to solve the inverse problem (2)–(1), we will reformulate it into an optimal control problem.

2.1 Formulation to Optimal Control Problem

One of the routinely used approach to study the inverse problem given by the Eqs. (2)–(1), is its formulation by an approximate optimal control problem.

Consider the set of admissible functions $\mathcal{U}_{ad}$ defined by

$$\mathcal{U}_{ad} = \{ f \in L^2(\Omega) \mid 0 \leq \alpha \leq f \leq \beta \ pp \ \Omega \ \text{ and } \ \|\nabla f\|_{0,\Omega} \leq c \}$$

where α, β and c are are given strictly positive real numbers. Thus, we can reformulate the inverse problem (2)–(1) into an optimization problem in the form

$$\begin{cases} find \ f^* \in \mathcal{U}_{ad} \ \text{ such that } \ : \\ J(f^*) = \inf\limits_{f \in \mathcal{U}_{ad}} J(f) := \frac{1}{2} \int_{\Gamma_0} (u - \varphi)^2 \, d\sigma, \\ \text{where } u \text{ is solution of } \begin{cases} -\Delta u + pu = f \ dans \ \Omega, \\ \partial_n u = g \ sur \ \ \Gamma. \end{cases} \end{cases} \tag{3}$$

We prove that the problem (3) is equivalent to the inverse problem (2)–(1) under certain assumptions.

Lemma 1 *Suppose that the problem (2)–(1) admits a solution (f_0, u_0) such that $f_0 \in \mathcal{U}_{ad}$, then the problem (2)–(1) is equivalent to (3).*

Proof Let f^* be a solution of (3), hence $f^* \in \mathcal{U}_{ad}$ and $J(f^*) \leq J(f) \ \ forall f \in \mathcal{U}_{ad}$ and $u^* = u(f^*)$ is solution of

$$\begin{cases} -\Delta u^* + pu^* = f^* \ in \ \Omega, \\ \partial_n u^* = g \ \ on \ \Gamma. \end{cases} \tag{4}$$

To show that (f^*, u^*) is a solution of (2)–(1), it remains to verify that $u^* = \varphi$ on Γ_0. Since $J(f^*) \le J(f)$ $\forall f \in \mathcal{U}_{ad}$, in particular $0 \le J(f^*) \le J(f_0) = \frac{1}{2}\int_{\Gamma_0}(u_0 - \varphi)^2\, d\sigma$, or (f_0, u_0) is solution of (2)–(1), so $u_0 = \varphi$ pp Γ_0 So $0 \le J(f^*) = \frac{1}{2}\int_{\Gamma_0}(u^* - \varphi)^2\, d\sigma \le 0$.
Hence $u^* = \varphi$ pp on Γ_0. Hence (f^*, u^*) is a solution of (2)–(1).
Conversely, we have (f_0, u_0) is solution of (2)–(1) with $f_0 \in \mathcal{U}_{ad}$, i.e.

$$\begin{cases} -\Delta u_0 + p u_0 = f_0 \; in \; \Omega, \\ \partial_n u_0 = g \; on \; \Gamma, \end{cases} \tag{5}$$

and $u_0 = \varphi$ on Γ_0. So $J(f_0) = \frac{1}{2}\int_{\Gamma_0}(u_0 - \varphi)^2\, d\sigma = 0 \le J(f)$ $\forall f \in \mathcal{U}_{ad}$. So f_0 is solution of (3).

Remark 1 In the case where the hypothesis of the previous lemma is not verified, the optimization problem (3) presents an efficient way to approximate the inverse problem (2)–(1).

In the following, we will study the existence of an optimal solution of (3).

3 Existence of an Optimal Solution

In order to show the existence of an optimal solution we first show that the optimization problem (3) is well-posed. Namely, we show that given $f \in \mathcal{U}_{ad}$ the problem state (2) admits a unique solution.

3.1 Existence and Uniqueness of the Solution of the State Problem

We write the variational formulation of the problem (2):

$$\begin{cases} \text{Find } u \in H^1(\Omega) \text{ such as} \\ \displaystyle\int_\Omega \nabla u \, \nabla v \, dx + \int_\Omega p\, u\, v\, dx = \int_\Omega f\, v\, dx + \int_\Gamma g\, v\, d\sigma, \quad \forall v \in H^1(\Omega). \end{cases} \tag{6}$$

Proposition 1 *The problem (6) admits a unique solution.*

Proof The proof of this result follows from the Lax-Milgrame lemma. Indeed, let $f \in \mathcal{U}_{ad}$, we define the forms a and L by:

$$a(u, v) = \int_\Omega \nabla u\, \nabla v\, dx + \int_\Omega p\, u\, v\, dx$$

and

$$L(v) = \int_{\Omega} f\, v\, dx + \int_{\Gamma} g\, v\, d\sigma.$$

The form a is bilinear and the form L is linear by linearity of the integral. Let u and v in $H^1(\Omega)$

$$\begin{aligned} a(u,u) &= \int_{\Omega} (\nabla u)^2 dx + \int_{\Omega} p(u)^2 \\ &\geq \|\nabla u\|_{0,\Omega}^2 + c_0\|u\|_{0,\Omega}^2 \\ &\geq min(1, c_0)\|u\|_{1,\Omega}^2. \end{aligned}$$

$$\begin{aligned} |a(u,v)| &\leq \int_{\Omega} |\nabla u||\nabla v| dx + \|p\|_{\infty} \int_{\Omega} |u||v| dx \\ &\leq \|\nabla u\|_{0,\Omega}^2 \|\nabla v\|_{0,\Omega}^2 + \|p\|_{\infty} \|u\|_{0,\Omega}^2 \|v\|_{0,\Omega}^2 \\ &\leq (1 + \|p\|_{\infty}) \|u\|_{1,\Omega}^2 \|v\|_{1,\Omega}^2. \end{aligned}$$

$$\begin{aligned} |L(v)| &\leq \int_{\Omega} |f||v| dx + \int_{\Gamma} |g||v| d\sigma \\ &\leq \|f\|_{0,\Omega}\|v\|_{0,\Omega} + \|g\|_{0,\Gamma}\|v\|_{0,\Gamma} \\ &\leq \left(\|f\|_{0,\Omega} + C\|g\|_{0,\Gamma}\right) \|v\|_{1,\Omega}. \end{aligned}$$

This follows from Hölder's inequality and the continuity of the trace map from $H^1(\Omega)$ into $L^2(\Gamma)$. Hence the continuity of the forms a and L and the coercivity of a.

In order to show the existence of a solution of (3), it suffices to show that $\mathcal{U}_{ad}$ is compact for an adequate topology on $\mathcal{U}_{ad}$ and the cost functional J is continuous on $\mathcal{U}_{ad}$. Then, since the functional J is composed of two functions as follows: $f \mapsto u(f) \mapsto J(f)$, its continuity follows from that of the state problem which results in the continuity of the map $f \mapsto u(f)$, then the continuity of the map J.

The following result prove the compacity of the space of admissible functions

Lemma 2 *The set $\mathcal{U}_{ad}$ is compact for the weak topology of $H^1(\Omega)$ and the strong topology of $L^2(\Omega)$.*

Proof Let $(f_k)_k$ be a sequence of elements of $\mathcal{U}_{ad}$, then $\mathcal{U}_{ad}$, then

$$f_k \in L^2(\Omega) \text{ and } \forall k,\ 0 \leq \alpha \leq f_k \leq \beta\ pp \text{ in } \Omega \text{ and } \|\nabla f\|_{0,\Omega} \leq c.$$

Then

$$\|f_k\|_{1,\Omega}^2 = \|f\|_{0,\Omega}^2 + \|\nabla f\|_{0,\Omega}^2 \leq \beta^2 mes(\Omega) + c,$$

thus

$$\| f_k \|_{1,\Omega}^2 \leq \sqrt{\beta^2 mes(\Omega) + c}.$$

Thus, using the weak compactness of the ball $B_f(0, c)$ in $H^1(\Omega)$, we can extract from the sequence $(f_k)_k$ a sub-sequence noted again $(f_k)_k$ such that

$$f_k \to f^* \text{ weakly in } H^1(\Omega).$$

On the other hand by using the compact injection of $H^1(\Omega)$ into $L^2(\Omega)$, we can extract a sub-sequence denoted again $(f_k)_k$ such that

$$f_k \to f^* \text{ in } L^2(\Omega).$$

Let us show that $f^* \in \mathcal{U}_{ad}$. Since we have

$$f_k \rightharpoonup f^* \text{ in } H^1(\Omega) \text{ and } \nabla : H^1(\Omega) \to L^2(\Omega) \text{ is weakly continuous.}$$

Hence

$$\nabla f_k \rightharpoonup \nabla f^* \text{ in } L^2(\Omega).$$

Or

$$\| \nabla f_k \|_{0,\Omega} \leq c \quad \forall\, k,$$

then $\nabla f_k \in B_f(0, c)$ which is weakly compact in $L^2(\Omega)$. Or this one is reflexive, so $\nabla f^* \in B_f(0, c)$. Consequently we have

$$\| \nabla f^* \|_{0,\Omega} \leq c.$$

On the other hand, we have

$$0 \leq \alpha \leq f_k \leq \beta \; pp \; \Omega \; \forall\, k.$$

Then

$$\alpha \leq f_k(x) \leq \beta \; \forall x \in \Omega \setminus M_k, \;\; mes(M_k) = 0 \quad \forall\, k,$$

or as

$$f_k \to f^* \text{ in } L^2(\Omega)$$

one can extract a sub-sequence denoted again $(f_k)_k$ such that

$$f_k \to f^*; \; pp \text{ in } \Omega$$

so

$$\alpha \leq f_k(x) \leq \beta \; \forall \; x \in \Omega \setminus M$$

with

$$M = \cup_{k\in\mathbb{N}} M_k \text{ and } mes(M) = 0.$$

This implies that $\alpha \le f^*(x) \le \beta \quad \forall\, x \in (\Omega \setminus M)$ with $mes(M) = 0$. Hence

$$\alpha \le f^* \le \beta \quad pp \text{ in } \Omega.$$

Therefore $f^* \in \mathcal{U}_{ad}$. Consequently, $\mathcal{U}_{ad}$ is compact for the weak topology of $H^1(\Omega)$ and strong topology of $L^2(\Omega)$.

We are going to show that if $(f_k)_k$ is a sequence of elements of $\mathcal{U}_{ad}$ which converges to f^* weakly in $H^1(\Omega)$ (or strongly in $L^2(\Omega)$) $u_k = u(f_k)$ the solution of (6) for $f = f_k$ converges to $u^* = u(f^*)$ the solution of (6) for $f = f^*$ in a certain sense that we will specify.

Proposition 2 *Let $(f_k)_k$ be a sequence of $\mathcal{U}_{ad}$ which converges to f^* in $\mathcal{U}_{ad}$, then $u_k = u(f_k)$ the solution of (6) for $f = f_k$ satisfies:*

(i) There exists $c > 0$ independent of k, such that $\|u_k\|_{1,\Omega} \le c$.
(ii) We can extract a subsequence denoted again by $(u_k)_k$ which converges weakly to $u^* = u(f^*)$ in $H^1(\Omega)$ or u^* is solution of (6) for $f = f^*$.

Proof (i) Denote by $u_k = u(f_k) \in H^1(\Omega)$ the solution of variational problem

$$\int_\Omega \nabla u_k \nabla v dx + \int_\Omega p u_k v dx = \int_\Omega f v dx + \int_\Gamma g v d\sigma \ \forall v \in H^1(\Omega)$$

in particular for $v = u_k$, we will have

$$\int_\Omega \nabla u_k \nabla u_k dx + \int_\Omega p(u_k)^2 dx = \int_\Omega f_k u_k dx + \int_\Gamma g u_k d\sigma$$

using the coercivity of the bilinear form and the continuity of the linear form, $\exists \gamma > 0$ and $M > 0$ independent of k, such that:

$$\gamma \|u_k\|^2_{1,\Omega} \le M \|u_k\|_{1,\Omega}.$$

Hence,

$$\|u_k\|_{1,\Omega} \le \frac{M}{\gamma} = c.$$

(ii) From this inequality, there exists a subsequence denoted again by (u_k), such that $u_k \rightharpoonup w$ in $H^1(\Omega)$, thanks to the weak compactness of $B_f(0, c)$ in $H^1(\Omega)$ which is reflexive, it then remains to show that $w = u(f^*)$ the solution of (6) for $f = f^*$ or we have $u_k \in H^1(\Omega)$ is solution of

$$\int_\Omega \nabla u_k \nabla v dx + \int_\Omega p u_k v dx = \int_\Omega f_k v dx + \int_\Gamma g v d\sigma.$$

Passing to the limit in this equation if we show that

$$\int_\Omega \nabla u_k \nabla v dx \to \int_\Omega \nabla w \nabla v dx$$

$$\int_\Omega p u_k v dx \to \int_\Omega p w v dx$$

and

$$\int_\Omega f_k v dx \to \int_\Omega f^* v dx.$$

Thus we will have shown that w is a solution of

$$\int_\Omega \nabla w \nabla v dx + \int_\Omega p w v dx = \int_\Omega f^* v dx + \int_\Gamma g v d\sigma \; \forall v \in H^1(\Omega)$$

which admits a unique solution $u(f^*)$ and therefore $w = u(f^*)$.
Indeed we have

$$u_k \rightharpoonup w \text{ in } H^1(\Omega) \text{ and } \nabla : H^1(\Omega) \to L^2(\Omega)$$

is linear and therefore continuous

$$\nabla u_k \to \nabla w \text{ in } L^2(\Omega)$$

i.e

$$\int_\Omega (\nabla u_k - \nabla w) v dx \to 0 \; \forall v \in L^2(\Omega).$$

On the other hand, we have

$$|\int_\Omega p(u_k - w) v dx| \leq \|p\|_\infty \|u_k - w\|_{0,\Omega} \|v\|_{0,\Omega}.$$

But $u_k \rightharpoonup w$ in $H^1(\Omega)$, according to the compact injection of $H^1(\Omega)$ in $L^2(\Omega)$, we will have

$$u_k \to w \text{ in } L^2(\Omega),$$

so

$$\int_\Omega p(u_k - w) v dx \to 0.$$

On the other hand, we have

$$|\int_{\Omega}(f_k - f^*)v dx| \leq \|f_k - f^*\|_{0,\Omega}\|v\|_{0,\Omega}.$$

As

$$f_k \to f^* \text{ in } L^2(\Omega),$$

we get the result.

3.2 Continuity of Cost Functional

We have the following continuity result:

Proposition 3 *Let $(p_k)_k$ be a sequence of elements of $\mathcal{U}_{ad}$ which converges to p^* in $\mathcal{U}_{ad}$, then*

$$\lim_{k\to+\infty} J(f_k) = J(f^*).$$

Proof Let $(f_k)_k$ be a sequence of elements of $\mathcal{U}_{ad}$ which converges to f^*, we have already shown that the sequence $(u_k)_k$ such that $u_k = u(f_k)$ is the solution of (6) for $f = f_k$ converges weakly in $H^1(\Omega)$ to $u^* = u(f^*)$ solution of (6) for $f = f^*$. We have :

$$\begin{aligned} J(f_k) - J(f^*) &= \int_{\Gamma_0}(u_k - \varphi)^2 d\sigma - \int_{\Gamma_0}(u^* - \varphi)^2 d\sigma \\ &= \int_{\Gamma_0}(u_k - \varphi + u^* - \varphi)(u_k - \varphi - u^* + \varphi)d\sigma \\ &= \int_{\Gamma_0}(u_k + u^* - 2\varphi)(u_k - u^*)d\sigma. \end{aligned}$$

Using Hölder's inequality we get

$$\begin{aligned} |J(f_k) - J(f^*)| &\leq \int_{\Gamma_0}|u_k + u^* - 2\varphi||u_k - u^*|d\sigma \\ &\leq \|u_k + u^* - 2\varphi\|_{L^2(\Gamma_0)}\|u_k - u^*\|_{L^2(\Gamma_0)} \\ &\leq (\|u_k\|_{L^2(\Gamma_0)} + \|u^*\|_{L^2(\Gamma_0)} + 2\|\varphi\|_{L^2(\Gamma_0)})\|u_k - u^*\|L^2(\Gamma_0). \end{aligned}$$

Using the continuity of the trace map of $H^1(\Omega)$ in $L^2(\Gamma_0)$, $\exists M > 0$ such that

$$\|u_k\|_{L^2(\Gamma_0)} \leq M\|u_k\|_{1,\Omega} \leq Mc \quad \forall\, k \in \mathbb{N}.$$

On the other hand, using the compactness of the trace map of $H^1(\Omega)$ in $L^2(\Gamma_0)$, since $u_k \to u^*$ in $H^1(\Omega)$ we can extract a subsequence noted again $(u_k)_k$ such that $u_k|_{\Gamma_0} \to u^*|_{\Gamma_0}$ in $L^2(\Gamma_0)$ i.e $\|u_k - u^*\|_{L^2(\Gamma_0)} \to 0$ which allows us to conclude.

Theorem 1 *The optimization problem (3) admits at least one solution in $\mathcal{U}_{ad}$.*

Proof Let (f_k) be a minimizing sequence of J on $\mathcal{U}_{ad}$ i.e $J(f_k) \to \inf_{f \in \mathcal{U}_{ad}} J(f)$. By compactness of $\mathcal{U}_{ad}$ there exists a subsequence denoted again (f_k) which converges in $\mathcal{U}_{ad}$ to f^*, or J is continuous on $\mathcal{U}_{ad}$, then $J(f_k) \to J(f^*)$, so $J(f^*) = \inf_{f \in \mathcal{U}_{ad}} J(f)$.

Remark 2 Note that the uniqueness of the solution of an optimization problem stems from the convexity of the functional cost. Actually, this condition is not easy to obtain in the case of the functional J.

4 Numerical Reconstruction Method

In this section, we deal with the problem of optimization (3) by the method of the gradient type, for that, we will need the gradient of the functional J with respect to the second member. First, we establish the differentiability of the solution $u(f)$ with respect to f.

4.1 State Sensitivity Study

We introduce the following sensitivity problem:

$$u^1 = u^1(f, \lambda) = \lim_{t \to 0} \frac{u(f + t\lambda) - u(f)}{t}$$

for all $f \in L^2(\Gamma_0)$ and $f \in \mathcal{U}_{ad}$, with $u(f + t\lambda)$ satisfies the state problem (2) replacing f by $f + t\lambda$, i.e.:

$$\begin{cases} -\Delta u(f + t\lambda) + pu(f + t\lambda) = (f + t\lambda) \; in \; \Omega, \\ \partial_n u(f + t\lambda) = \quad g \quad on \; \Gamma. \end{cases}$$

Let $w_t = \dfrac{u(f + t\lambda) - u(f)}{t}$, then w_t satisfies the problem:

$$\begin{cases} -\Delta w_t + pw_t = \lambda \; in \; \Omega, \\ \partial_n w_t = 0 \; on \; \Gamma. \end{cases}$$

When $t \to 0$, we get the following sensitivity problem:

$$\begin{cases} -\Delta u^1 + pu^1 = \lambda \; in \; \Omega, \\ \partial_n u^1 = 0 \; on \; \Gamma. \end{cases} \tag{7}$$

4.2 Gradient Calculation

Before calculating the gradient of our cost functional, we first look for the adjoint equation associated with our problem.

The augmented Lagrangian associated with the problem (3) is written:

$$\begin{aligned}\mathcal{L}(u, v, f) &= J(f) - a(u, v) + L(v)\\ &= \frac{1}{2}\int_{\Gamma_0} (u-\varphi)^2 \, d\sigma - \int_{\Omega} \nabla u \, \nabla v \, dx - \int_{\Omega} p \, u \, v \, dx + \int_{\Omega} f \, v \, dx + \int_{\Gamma} g \, v \, d\sigma.\end{aligned}$$

The adjoint state is solution of $\dfrac{d\mathcal{L}}{du} = 0$:

$$\begin{aligned}\frac{d\mathcal{L}(u, v, f)}{du}(\psi) &= \lim_{t\to 0} \frac{\mathcal{L}(u+t\psi, v, f) - \mathcal{L}(u, v, f)}{t}\\ &= \int_{\Gamma_0} (u-\varphi)\,\psi \, d\sigma - \int_{\Omega} \nabla\psi \, \nabla v \, dx - \int_{\Omega} p \, \psi \, v \, dx.\end{aligned}$$

Using Green's formula we get:

$$\frac{d\mathcal{L}(u, v, f)}{du}(\psi) = \int_{\Gamma_0} (u-\varphi)\,\psi \, d\sigma + \int_{\Omega} \Delta v \psi \ dx - \int_{\Gamma} \partial_n v \psi \, d\sigma - \int_{\Omega} p \, v \, \psi \, dx.$$

This implies that v is the solution of the following conjoint state:
Find $v(f) \in H^1(\Omega)$ such that

$$\begin{cases} -\Delta v + pv = \quad 0 & in \ \Omega, \\ \partial_n v = u(f) - \varphi & on \ \Gamma_0, \\ \partial_n v = \quad 0 & on \ \Gamma_1. \end{cases} \tag{8}$$

Theorem 2 *The functional J is Fréchet differentiable, and its derivative $J'(f)$ at $f \in \mathcal{U}_{ad}$ is given by:*

$$I(f)[\lambda] = \int_{\Omega} \lambda v(f) \, dx \tag{9}$$

where $v(f)$ is the solution of the adjoining state problem (8).

Proof The differential of J is given by:

$$J'(f)[\lambda] = \int_{\Gamma_0} (u(f) - \varphi) u^1(f, \lambda) ds.$$

By writing the variational formulation of the adjoint state problem (8) with u^1 as the test function, and the variational formulation of the sensitivity problem (7) with v as the test function , we get the following formulas:

$$\int_{\Omega} \nabla v(f)\nabla u^1(f,\lambda) + pv(f)u^1(f,\lambda)\,dx - \int_{\Gamma_0} (u(f)-\varphi)u^1(f,\lambda)\,ds = 0$$
$$\int_{\Omega} \nabla u^1(f,\lambda)\nabla v(f) + pu^1(f,\lambda)v(f)\,dx - \int_{\Omega} \lambda v(f)\,dx = 0.$$

Which implies

$$\int_{\Gamma_0} (u(f)-\varphi)u^1(f,\lambda)\,ds = \int_{\Omega} \lambda v(f)\,dx,$$

Consequently

$$I(f)[\lambda] = \int_{\Omega} \lambda v(f)\,dx.$$

4.3 Approximation in a Reproducing Kernel Space

In this section, we will restrict the search space of the source f to a reproducing kernel Hilbert space. We assume that $H = \overline{span\{K_x \,/\, x \in \Omega\}}$.

In order to guarantee the stability of the problem during the numerical resolution, and to avoid that a small disturbance on the data causes a large error in the solution, we use the Tikhonov regularization. It consists of adding the regularization term $\nu\|f\|_H^2$ to the functional to be minimized. We therefore come back to the problem

$$\inf_{f\in H} \left\{J(f) + \nu\|f\|_H^2\right\}$$

with $\|.\|_H$ is the kernel-space norm reproducing H, and ν is a positive regularization parameter.

Then, by the representation theorem [33], for all $x \in \Omega$ we have

$$f(x) = \sum_{i=1}^{d} \alpha_i K(x, x_i).$$

So, instead of finding $f \in H$ the minimum of the functional $J(f)$, we are reduced to finding of the vector $\alpha = (\alpha_1, \alpha_2, \ldots, \alpha_d) \in \mathbb{R}^d$ minimizing the functional

$$\tilde{J}(\alpha) = J(f) = J\left(\sum_{i=1}^{d} \alpha_i K(., x_i)\right) = \|\sum_{i=1}^{d} \alpha_i K(., x_i) - \varphi\|_H^2$$

$$\inf_{\alpha\in\mathbb{R}^d} \mathcal{J}(\alpha) := \tilde{J}(\alpha) + \nu \sum_{k=1}^{d}\sum_{i=1}^{d} \alpha_i\alpha_k K(x_k, x_i) \tag{10}$$

Here we note that $\mathcal{J}$ is a convex function.

4.4 Numerical Algorithm

The conjugate gradient algorithm applied to the considered optimization problem takes the following form:

Algorithm 1: Gradient algorithm

1: Choose an initial vector $\alpha_0 = (\alpha_1, ..., \alpha_N) \in \mathbb{R}^N$ and put $k = 0$ ($f_0 = \sum_{i=1}^{N} \alpha_0^i K_{x_i}$).
2: Solve the direct problem and the adjoint problem for $f = f_k = \sum_{i=1}^{N} \alpha_k^i K_{x_i}$.
3: Determine the gradient $\nabla J(\alpha_k)$ using the formula (9) and the direction of descent $d_k = -\nabla J(\alpha_k)$.
4: Update vector α_k by $\alpha_{k+1} = \alpha_k + \beta d_k$
5: End.

5 Numerical Results

In this section we will consider some examples with synthetic data in order to show the effectiveness of our approach. Consider the problem (2)–(1) with the following data:

$$\Omega =]0, 1[\times]0, 1[, \quad p = 1, \quad g = 0.$$

We are going to apply the Algorithm 1. The state problem is approximated by the finite element method.

Example 1 In this first example we assume that the original source term $f \in H$, i.e. that

$$f = \sum_{i=1}^{d} \alpha_i K_{x_i},$$

with $\alpha_i = 1, \quad i = 1, \dots d$.
The problem (2)–(1) therefore amounts to looking for $(\alpha_1, \dots, \alpha_d) \in \mathbb{R}^d$.

In the Figs. 1 and 2 we present the results obtained by applying the Algorithm 1. As can be seen, the results obtained are of good quality.

Example 2 In this second example we assume that the original source term is not necessarily in the space H. We consider f under the following form

$$f = \sin(2\pi x)\cos(y).$$

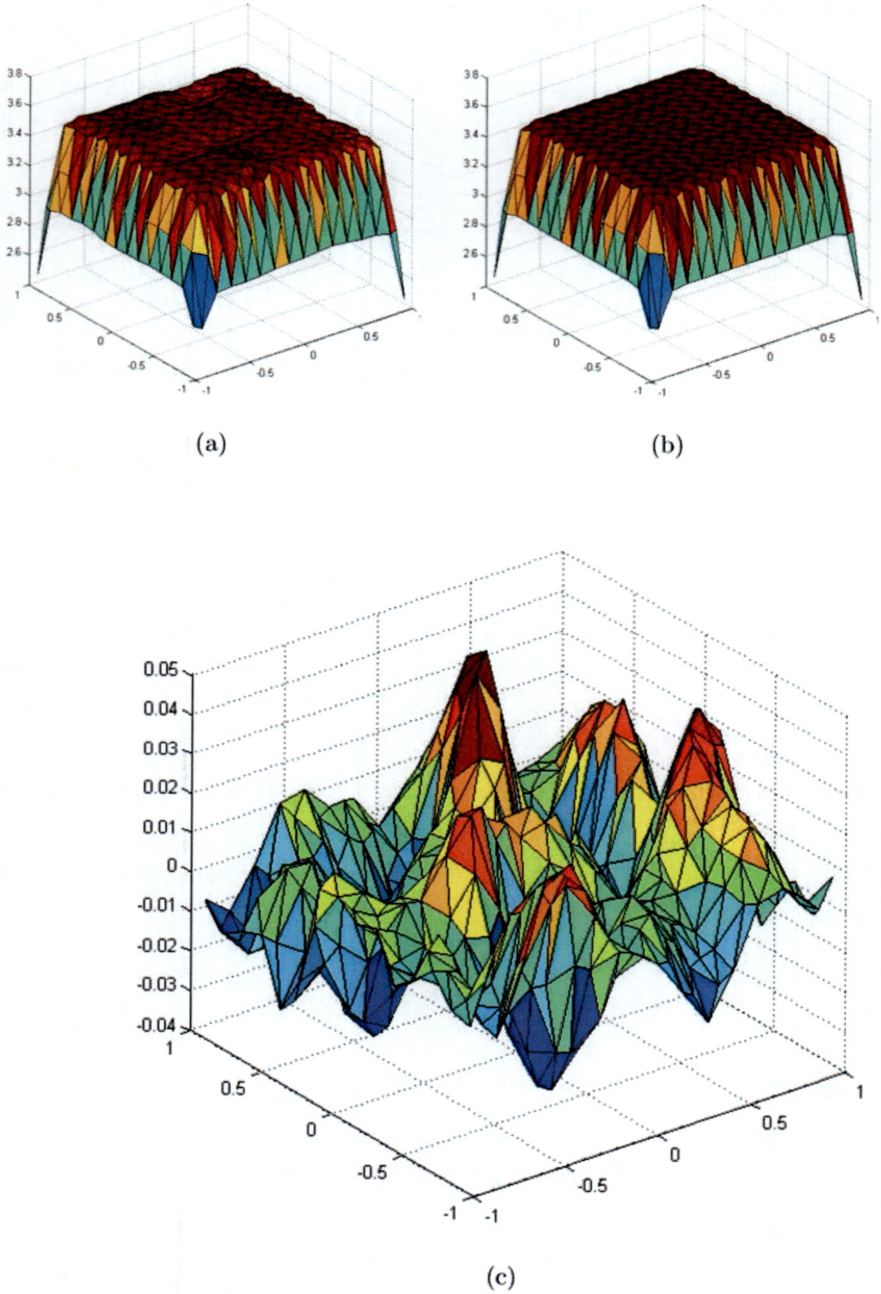

Fig. 1 **a** The optimal source term. **b** The exact source term. **c** The difference between optimal and exact source term

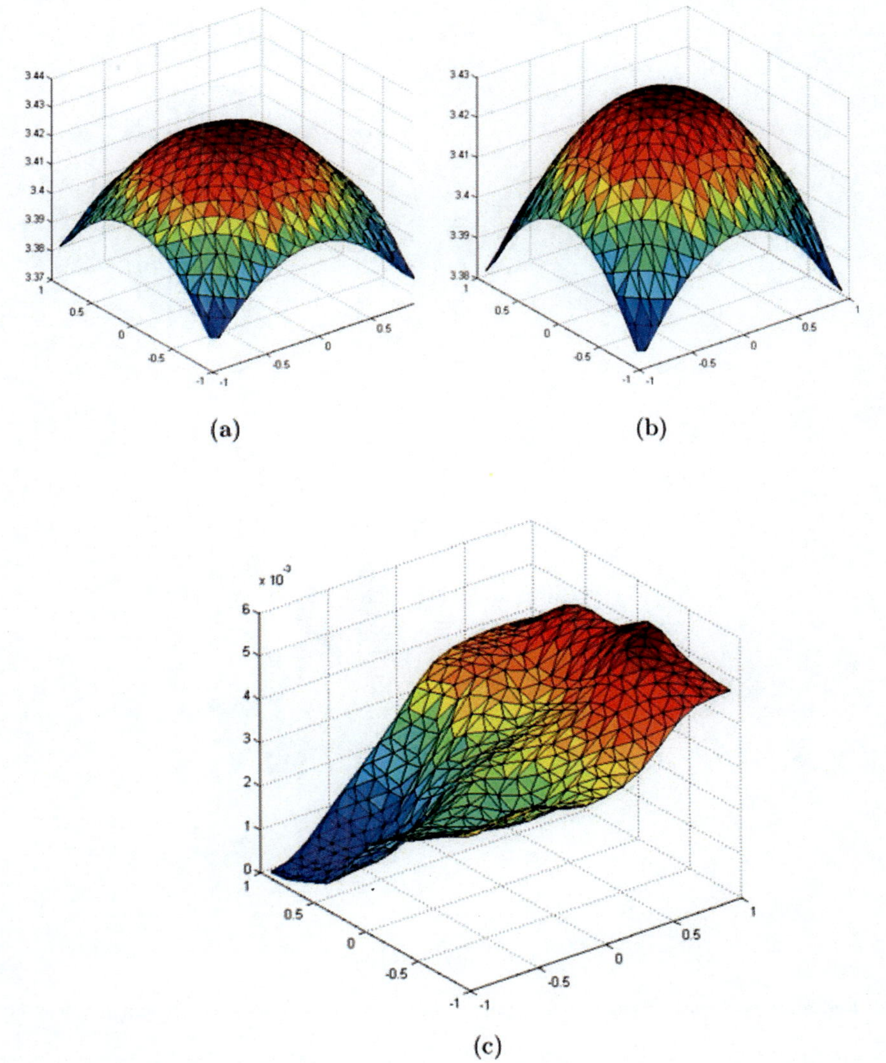

Fig. 2 The comparison between the optimal and exact solution is represented in **a**, **b** and **c**

The problem (2)–(1) therefore amounts to looking for $(\alpha_1, \ldots, \alpha_d) \in \mathbb{R}^d$, such that $\sum_{i=1}^{d} \alpha_i K_{x_i}$ approaches f. In the Figs. 3 and 4, we present the obtained results.

We notice that in this second although the quality of approximation is slightly less good than Example 1, the results remain very satisfactory.

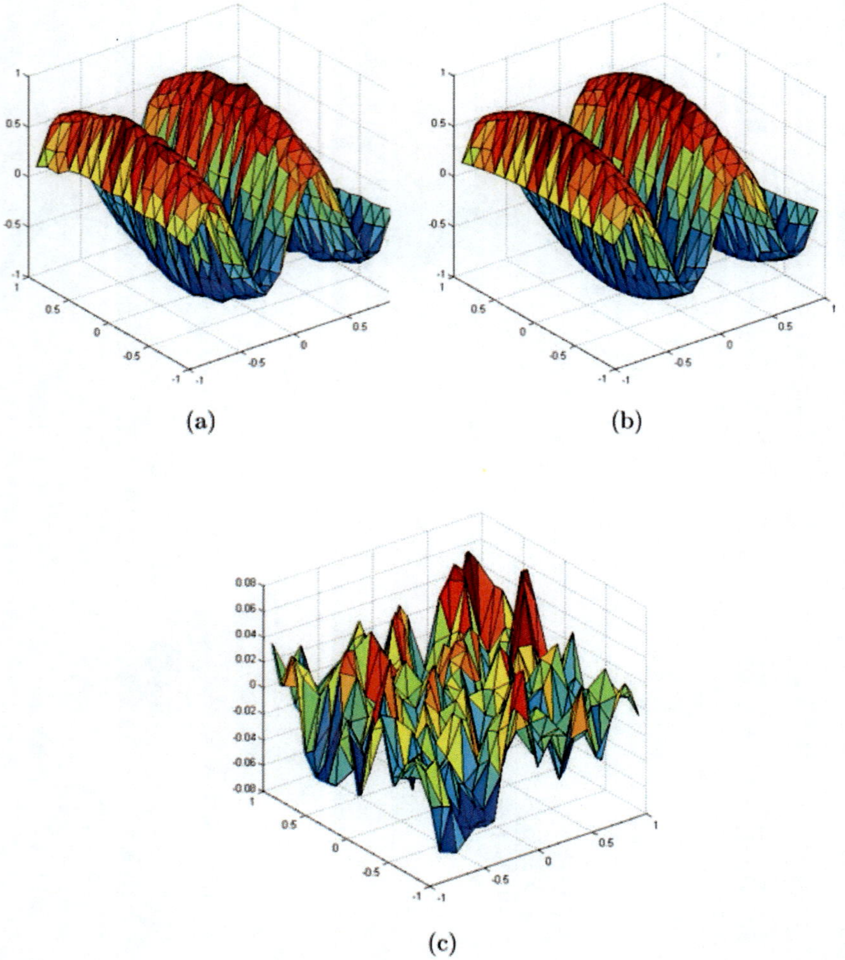

Fig. 3 The comparison between the optimal and exact source term is represented in **a**, **b** and **c**

6 Conclusion

In this work, we have dealt with an inverse problem of term source identification. In order to solve this problem, an optimal control formulation is proposed. Thus, we have shown an existence result of an optimal solution.

We applied a method to solve the obtained optimization problem based on the reproducing kernel Hilbert spaces. Then using a discretization based on the finite element method, the problem is reduced to a linear system. Thus, we have proposed an optimization method based on gradient descent algorithm. Finally, we presented numerical results that prove the effectiveness of the proposed approach.

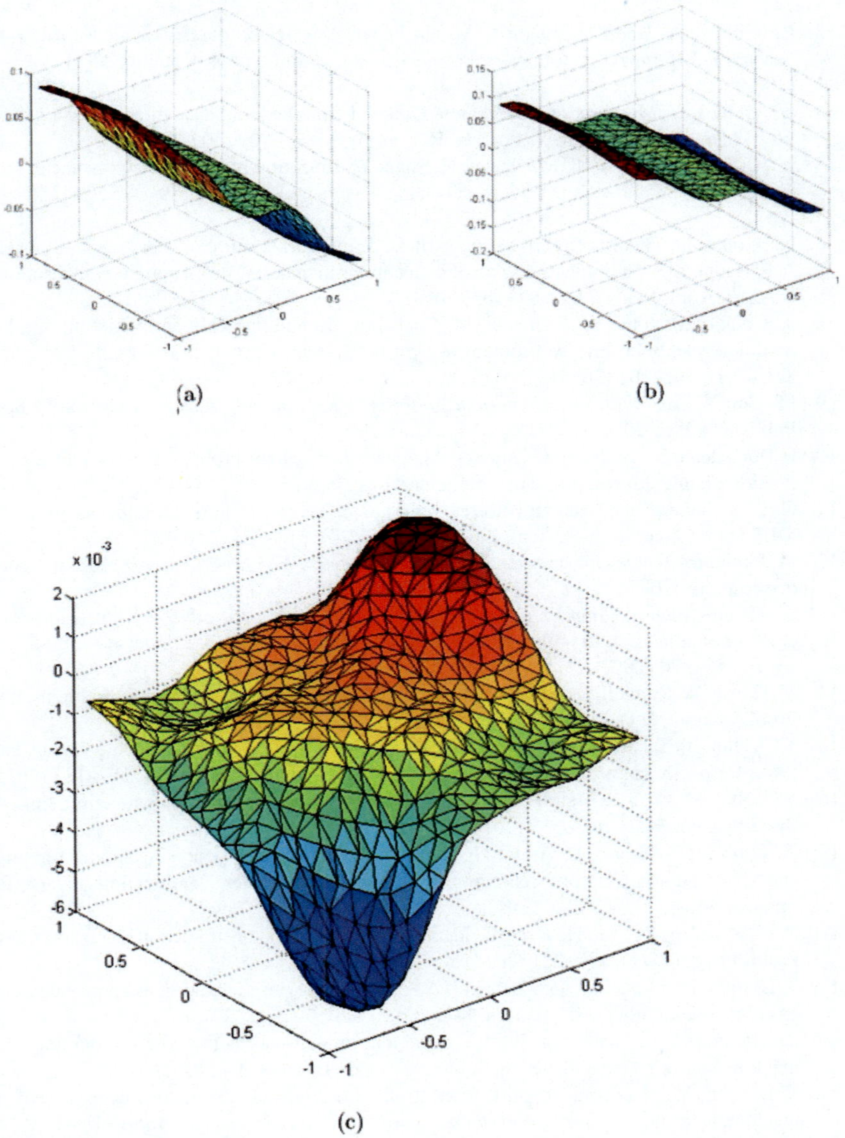

Fig. 4 The comparison between the optimal and exact solution is represented in **a**, **b** and **c**

References

1. B. Abdelaziz, A.E.L. Badia, A.E.L. Hajj, Identification of pointwise sources in a bioluminescent tomography problem, in *ESAIM: Proceedings*, vol. 45 (2014), pp. 390–399
2. L. Afraites, A. Bellouquid, Global optimization approaches to parameters identification in immune competition model. Commun. Appl. Ind. Math. **5** (2014)

3. L. Afraites, A. Hadri, A. Laghrib, M. Nachaoui, A weighted parameter identification PDE-constrained optimization for inverse image denoising problem. Vis. Comput. **38**(8), 2883–2898 (2022)
4. L. Afraites, C. Masnaoui, M. Nachaoui, Coupled complex boundary method for a geometric inverse source problem. RAIRO-Oper. Res. **56**(5), 3689–3709 (2022)
5. L. Afraites, C. Masnaoui, M. Nachaoui, Shape optimization method for an inverse geometric source problem and stability at critical shape. Discrete Contin. Dyn. Syst. Ser. S **15**(1), 1–21 (2022)
6. N. Aronszajn, Theory of reproducing kernels. Trans. Am. Math. Soc. **68**(3), 337–404 (1950)
7. A.E. Badia, A.E. Hajj, Identification of dislocations in materials from boundary measurements. SIAM J. Appl. Math. **73**(1), 84–103 (2013)
8. A. Chakib, M. Johri, A. Nachaoui, M. Nachaoui, On a numerical approximation of a highly nonlinear parabolic inverse problem in hydrology. Ann. Univ. Craiova. Math. Comput. Sci. Ser. **42**(1), 192–201 (2015)
9. M. Cui, Y. Lin, *Nonlinear Numerical Analysis in Reproducing Kernel Space* (Nova Science Publishers, Inc., 2009)
10. P. Duchateau, R. Thelwell, G. Butters, Analysis of an adjoint problem approach to the identification of an unknown diffusion coefficient. Inverse Probl. **20**(2), 601 (2004)
11. G.E. Fasshauer, Dual bases and discrete reproducing kernels: a unified framework for RBF and MLS approximation. Eng. Anal. Bound. Elem. **29**(4), 313–325 (2005)
12. M. Farah, *Problémes Inverses de Sources et Lien avec l'Electro-encéphalo-graphie* (Doctoral dissertation) (2007)
13. A. Hamdi, Identification of point sources in two-dimensional advection-diffusion-reaction equation: application to pollution sources in a river. Stationary case. Inverse Probl. Sci. Eng. **15**(8), 855–870 (2007)
14. M. Hanke, W. Rundell, On rational approximation methods for inverse source problems. Inverse Probl. Imaging **5**(1), 185–202 (2011)
15. Y.C. Hon, T. Takeuchi, Discretized Tikhonov regularization by reproducing kernel Hilbert space for backward heat conduction problem. Adv. Comput. Math. **34**(2), 167–183 (2011)
16. Y.C. Hon, M. Li, Y.A. Melnikov, Inverse source identification by Green's function. Eng. Anal. Bound. Elem. **34**(4), 352–358 (2010)
17. A. Lekbir, H. Aissam, L. Amine, N. Mourad, A non-convex denoising model for impulse and Gaussian noise mixture removing using bi-level parameter identification. Inverse Probl. Imaging **16**(4), 827–870 (2022)
18. Y. Li, S. Osher, R. Tsai, Heat source identification based on constrained minimization. Inverse Probl. Imaging **8**(1), 199–221 (2014)
19. S. Lyaqini, M. Quafafou, M. Nachaoui, A. Chakib, Supervised learning as an inverse problem based on non-smooth loss function. Knowl. Inf. Syst. 1–20 (2020)
20. S. Lyaqini, M. Nachaoui, A. Hadri, An efficient primal-dual method for solving non-smooth machine learning problem. Chaos Solitons Fractals **155**, 111754 (2022)
21. S. Lyaqini, M. Nachaoui, Identification of genuine from fake banknotes using an enhanced machine learning approach, in *Mathematical Control and Numerical Applications: JANO13*, Khouribga, Morocco, Feb 22–24, 2021 (Springer, 2021), pp. 59–70
22. M. Lukić, J. Beder, Stochastic processes with sample paths in reproducing kernel Hilbert spaces. Trans. Am. Math. Soc. **353**(10), 3945–3969 (2001)
23. M. Nachaoui, Parameter learning for combined first and second order total variation for image reconstruction. Adv. Math. Models Appl. **5**(1), 53–69 (2020)
24. M. Nachaoui, A. Nachaoui, T. Tadumadze, On the numerical approximation of some inverse problems governed by nonlinear delay differential equation. RAIRO-Oper. Res. **56**(3), 1553–1569 (2022)
25. M. Nachaoui, A. Nachaoui, T. Tadumadze, On the numerical approximation of some inverse problems governed by nonlinear delay differential equation. RAIRO-Oper. Res. **56**(3), 1553–1569 (2022)

26. M. Nachaoui, Étude théorique et approximation numérique d'un problème inverse de transfert de la chaleur (Doctoral dissertation, Université de Nantes) (2011)
27. M. Nachaoui, A. Laghrib, An improved bilevel optimization approach for image super-resolution based on a fractional diffusion tensor. J. Frankl. Inst. **359**(13), 7165–7195 (2022)
28. A. Oulmelk, L. Afraites, A. Hadri, M. Nachaoui, An optimal control approach for determining the source term in fractional diffusion equation by different cost functionals. Appl. Numer. Math. **181**, 647–664 (2022)
29. V.I. Paulsen, M. Raghupathi, *An Introduction to the Theory of Reproducing Kernel Hilbert Spaces*, vol. 152 (Cambridge University Press, 2016)
30. L.G. Olson, R.D. Throne, Computational issues arising in multidimensional elliptic inverse problems: the inverse problem of electrocardiography. Eng. Comput. **12**(4), 343–356 (1995)
31. A.R. Paiva, I. Park, J.C. Principe, A reproducing kernel Hilbert space framework for spike train signal processing. Neural Comput. **21**(2), 424–449 (2009)
32. S. Saitoh, *Theory of Reproducing Kernels and Its Applications* (Longman Scientific & Technical, 1988)
33. B. Schölkopf, R. Herbrich, A.J. Smola, A generalized representer theorem, in *International Conference on Computational Learning Theory* (Springer, Berlin, Heidelberg, 2001), pp. 416–426
34. A. Shaw, D. Roy, S.R. Reid, M. Aleyaasin, Reproducing kernel collocation method applied to the non-linear dynamics of pipe whip in a plane. Int. J. Impact Eng. **34**(10), 1637–1654 (2007)
35. T. Takeuchi, M. Yamamoto, Tikhonov regularization by a reproducing kernel Hilbert space for the Cauchy problem for an elliptic equation. SIAM J. Sci. Comput. **31**(1), 112–142 (2008)
36. M.H. Tber, M.E.A. Talibi, D. Ouazar, Parameters identification in a seawater intrusion model using adjoint sensitive method. Math. Comput. Simul. **77**(2–3), 301–312 (2008)
37. J.F. Wang, Reproducing kernel of image space of Haar wavelet transform, in *2008 International Conference on Wavelet Analysis and Pattern Recognition*, vol. 2 (IEEE, 2008), pp. 639–643
38. W. Wang, B. Han, M. Yamamoto, Inverse heat problem of determining time-dependent source parameter in reproducing kernel space. Nonlinear Anal. Real World Appl. **14**(1), 875–887 (2013)
39. C.G. Xanthis, P.M. Bonovas, G.A. Kyriacou, Inverse problem of ECG for different equivalent cardiac sources. Piers Online **3**(8), 1222–1227 (2007)
40. H. Xiang, B. Liu, Solving the inverse problem of an SIS epidemic reaction-diffusion model by optimal control methods. Comput. Math. Appl. **70**(5), 805–819 (2015)
41. S. Xu, X. Zhang, Determination of fracture parameters for crack propagation in concrete using an energy approach. Eng. Fract. Mech. **75**(15), 4292–4308 (2008)